# Internationalization, Technology and Services

PREST/CRIC STUDIES IN SCIENCE, TECHNOLOGY AND
INNOVATION

**Series Editors:** Luke Georghiou, *Director, PREST* and Stan Metcalfe, *Stanley Jevons Professor of Economics and ESRC Centre for Research on Innovation and Competition, University of Manchester, UK*

This series is a forum for the outstanding research from Manchester's two leading social science research institutes – PREST (Policy Research in Engineering, Science and Technology) and CRIC (the ESRC Centre for Research on Innovation and Competition).

The books in this series reflect their concern with the economics, social and managerial implications of science and innovation. Emphasis is given to science and technology policy and the role that innovation plays in competitiveness, particularly in the new knowledge-intensive global economy.

This interdisciplinary series will include some of the best theoretical and empirical work in the field, and will be invaluable to scholars, students and policy makers.

Titles in the series include:

Knowledge and Innovation in the New Service Economy
*Edited by Birgitte Andersen, Jeremy Howells, Richard Hull, Ian Miles and Joanne Roberts*

Technological Change and the Evolution of Corporate Innovation
The Structure of Patenting 1890–1990
*Birgitte Andersen*

Internationalization, Technology and Services
*Edited by Marcela Miozzo and Ian Miles*

# Internationalization, Technology and Services

*Edited by*

Marcela Miozzo

*Lecturer in Technology Management, Manchester School of Management, University of Manchester Institute of Science and Technology (UMIST), UK*

Ian Miles

*Co-director, PREST and CRIC and Professor of Technological Innovation and Social Change, University of Manchester, UK*

PREST/CRIC STUDIES IN SCIENCE, TECHNOLOGY AND INNOVATION

**Edward Elgar**
Cheltenham, UK • Northampton, MA, USA

Published by
Edward Elgar Publishing Limited
Glensanda House
Montpellier Parade
Cheltenham
Glos GL50 1UA
UK

Edward Elgar Publishing, Inc.
136 West Street
Suite 202
Northampton
Massachusetts 01060
USA

A catalogue record for this book
is available from the British Library

**Library of Congress Cataloguing in Publication Data**
Internationalization, technology, and services / edited by Marcela Miozzo,
 Ian Miles.
    p. cm. – (PREST/CRIC studies in science, technology, and innovation)
   Includes index.
   1. Technological innovations. 2. Globalization. I. Miozzo, Marcela, 1963-
II. Miles, Ian. III. Series.

HD45 .I565 2003
338.8′8–dc21                                                            2002029828

ISBN 1 84376 053 3

Printed and bound in Great Britain by MPG Books Ltd, Bodmin, Cornwall

# Contents

*List of figures*                                                                 vii
*List of tables*                                                                  viii
*List of contributors*                                                            xi

    Introduction                                              1
    *Marcela Miozzo and Ian Miles*

**PART I   INNOVATION AND INTERNATIONALIZATION OF
SERVICES: CONCEPTUAL ISSUES**

  1  The relation between the internationalization of services and
the process of innovation: a research agenda                                      15
*Marcela Miozzo and Ian Miles*

  2  Internationalization and the demarcation between services and
manufactures: a theoretical and empirical analysis                                33
*Grazia Ietto-Gillies*

**PART II   TECHNOLOGY AND TRADE AND FOREIGN
INVESTMENT IN SERVICES: A STATISTICAL
APPRAISAL**

  3  The internationalization of European services: what can data
on international services transactions tell us?                                    59
*Paul Baker, with Marcela Miozzo and Ian Miles*

  4  Internationalization of services: are the modes changing?          87
*Zbigniew Zimny and Padma Mallampally*

**PART III   INTERNATIONAL SERVICE MULTINATIONALS
AND THE LOCATION OF PRODUCTION AND
INNOVATION ACTIVITY**

  5  Globalization, regionalization and 'scales of integration': US
IT industry investment in Southeast Asia                                          117
*Neil M. Coe*

6    National versus international effects in regional concentration
     of European innovative business services                    137
     *Luis Rubalcaba-Bermejo and David Gago-Saldaña*

PART IV   INTERNATIONALIZATION AND INNOVATION:
          THE CHALLENGE FOR COUNTRIES AND REGIONS

7    From market to resource-oriented overseas expansion:
     re-examining a study of the internationalization of UK
     business service firms                                       161
     *Joanne Roberts*

8    Services internationalization: characteristics, potentials and
     barriers of Danish services firms                            184
     *Anders Henten and Torben Vad*

9    Internationalization of knowledge-intensive business services
     in a small European country: experiences from Finland        206
     *Marja Toivonen*

10   Services, scale and structures of internationalization:
     northwest England's environmental technologies firms          227
     *Sally Randles and Bruce Tether*

     *Index*                                                      255

# Figures

| | | |
|---|---|---:|
| 3.1 | Intra-EU trade and FDI flows in services relative to GDP | 76 |
| 3.2 | Extra-EU trade and FDI flows in services relative to GDP | 76 |
| 4.1 | Intra-firm exports of other private services, value and share in total exports, United States, 1986–2000 | 103 |
| 4.2 | Intra-firm imports of other private services, value and share in total imports, United States, 1986–2000 | 104 |
| 5.1 | Conceptualizing international expansion strategies | 120 |
| 5.2 | Typical organizational structures of US IT industry TNCs | 127 |
| 5.3 | Microsoft's global manufacturing and distribution operations | 131 |
| 6.1 | Business services and GDP per inhabitant: all regions and NUTS 1 | 153 |
| 7.1 | Other business service credits, by country, 1987–96 | 163 |
| 7.2 | Value of cross-border mergers and acquisitions in the business service sector, 1991–8 | 164 |
| 10.1 | Sales, by spatial reach amongst firms | 242 |

# Tables

| | | |
|---|---|---|
| 2.1 | World's largest 664 TNCs: breakdown by home country and sector, 1997 | 46 |
| 2.2 | World's largest 664 TNCs, by industry and home country, 1997 | 47 |
| 2.3 | World's largest 664 TNCs, by sector and industry: network of affiliates – indices and ranking, 1997 | 48 |
| 3.1 | Trade in service, by mode of supply, 1997 | 65 |
| 3.2 | Modes of supply and statistical data sources | 68 |
| 3.3 | Breakdown of services trade for Triad members in 1998 | 71 |
| 3.4 | Trade in services as a share of total trade by Triad members in 1998 | 71 |
| 3.5 | Annual average growth rates of trade for the Triad | 72 |
| 3.6 | The importance of trade in services for individual member states | 73 |
| 3.7 | The importance of FDI flows in services for individual member states | 74 |
| 3.8 | The importance of FDI stocks in services for individual member states | 75 |
| 3.9 | The share of services in intra-EU trade and FDI, and the intra-EU share in total services trade and FDI | 77 |
| 3.10 | Geographical composition of extra-EU transactions in service | 79 |
| 3.11 | Sector composition of extra-EU trade and FDI for services | 80 |
| 3.12 | Comparison of share of non-nationally owned enterprises in turnover and ratio of trade to turnover | 81 |
| 4.1 | Ratio of FDI sales to exports and imports, goods and services, United States, 1986–99 | 98 |
| 4.2 | Growth of international transactions in services by modes of delivery, United States, 1986–99 | 100 |
| 4.3 | Ratios of FDI sales to trade for selected services, United States, 1986–1999 | 101 |
| 4.4 | Imports of other private services into the United States, intra-firm and arm's-length, 1986 and 2000 | 106 |
| 4.5 | The importance of affiliate-to-affiliate trade in intra-firm trade, United States, 1977–99 | 108 |

4.6 Transactions of foreign affiliates in host countries, growth and composition 109

5.1 US IT industry FDI into Singapore 124

6.1 The importance of business services in European regions: breakdown by type of business services (five countries) 144

6.2 Explanatory factors by breakdown of business services categories: national v. international effects 147

6.3 Explanatory factors for advanced business services: national v. international effects 148

6.4 Correlations between business services and GDP per inhabitant 154

7.1 A classification of international activities conducted by business service firms 167

7.2 Factors as a source of competitive advantage 170

7.3 Factors that determine the location of an overseas presence 171

7.4 Examples of service activity located in the Philippines 174

8.1 Subsectors and number of respondents 185

8.2 Services or manufacturing 186

8.3 Modes and levels of internationalization 188

8.4 Size of firms and degree of internationalization 191

8.5 Size of firms and mode of internationalization 191

8.6 Degree of internationalization and attachment to manufactured goods 192

8.7 Degree of internationalization and personal contact 193

8.8 Degree of internationalization and need for local presence 194

8.9 Degree of internationalization and delivery via telecommunication lines 195

8.10 Degree of internationalization and standardization 196

8.11 Degree of internationalization and niche market orientation 196

8.12 Reasons of home market-oriented firms not to internationalize 197

8.13 Factors limiting international sales of service firms 199

9.1 Firms interviewed in different KIBS branches, by number of personnel 207

9.2 Market sector considered the most important, by business service branch, according to the Finnish SME barometer, 2001 210

9.3 Forms of internationalization in various KIBS branches in Finland 212

10.1 Environmental technologies and services in the Standard Industrial Classification 229

| | | |
|---|---|---|
| 10.2 | Firm size, by sales and employment | 235 |
| 10.3 | Firm size and specialization in environmental technologies and services | 236 |
| 10.4 | Firm size and the various firm types | 237 |
| 10.5 | Location of sales of ETS firms | 238 |
| 10.6 | Overseas markets | 238 |
| 10.7 | Logistic regressions for activity in overseas markets | 240 |
| 10.8 | Percentage of turnover from overseas markets | 243 |
| 10.9 | Modalities of international trade | 243 |
| 10.10 | Logistic regressions for modalities of international trade | 244 |

# Contributors

**Paul Baker**, Netherlands Economic Institute, Rotterdam, The Netherlands

**Neil M. Coe**, School of Geography, University of Manchester, Manchester, UK

**David Gago-Saldaña**, University of Alcala and Service Activities Research Laboratory (Servilab), Madrid, Spain

**Anders Henten**, Centre for Tele-Information, Technical University of Denmark, Lyngby, Denmark

**Grazia Ietto-Gillies**, The Business School, South Bank University, London, UK

**Padma Mallampally**, United Nations Conference on Trade and Development (UNCTAD), Geneva

**Ian Miles**, PREST and ESRC Centre for Research on Innovation and Competition (CRIC), University of Manchester, Manchester, UK

**Marcela Miozzo**, Manchester School of Management, UMIST, Manchester, UK

**Sally Randles**, ESCR Centre for Research on Innovation and Competition (CRIC), UMIST, Manchester, UK

**Joanne Roberts**, University of Durham Business School, Durham, UK

**Luis Rubalcaba-Bermejo**, University of Alcala and Service Activities Research Laboratory (Servilab), Madrid, Spain

**Bruce Tether,** ESCR Centre for Research on Innovation and Competition (CRIC), UMIST, Manchester, UK

**Marja Toivonen**, Employment and Economic Development Centre for Uusimaa, Helsinki, Finland

**Torben Vad**, PLS Rambøll Management, Copenhagen, Denmark

**Zbigniew Zimny**, United Nations Conference on Trade and Development (UNCTAD), Geneva

# Introduction

## Marcela Miozzo and Ian Miles

### BACKGROUND

This book is intended as a contribution to the emergent interdisciplinary debate on the internationalization of services. Most of the discussion has concerned public services and intensively traded mass services. In contrast, we are here paying particular attention to services that are knowledge- or technology-intensive. A clearer understanding of the process of internationalization of such services is required because it is an important part – arguably one of the most important parts – of the general process of globalization of production, distribution and innovation. It thus has implications for the international division of labour and the competitiveness of firms, regions and countries.

The growing importance of the internationalization of knowledge- or technology-intensive services has resulted from three main factors. With *economic development*, services have become major players in national economies. They naturally command more attention and strategic action, because of their significance to employment and economic growth. With *technological change*, services have become more important users of new technologies and players in innovation processes. There is a strong case that new technologies make the internationalization of some services more viable, in addition to the more conventional expectation that internationalization speeds up the pace of technological change in services. *With globalization*, the emergence of so-called 'global' firms has been as marked in services as elsewhere. Changes in the strategies of industrial multinational corporations have led to the increased internationalization of service firms, while changes in international policies through the European Single Market, the World Trade Organisation (WTO) and the Uruguay Round open the way to greater trade in services.

These three factors are closely intertwined. Moreover, the increased role and internationalization (through trade and foreign direct investment) of knowledge- or technology-intensive services is not only a product of these factors, but also a contributor to them. In turn, these factors are related to current developments in the world economy (and especially in the economies of advanced industrial societies). Among the most important of these are the following:

1. Substantial and long-term growth in the demand for knowledge- or technology-intensive services is reflected in the high growth rates of business services and a growing 'externalization' of such activities from firms in all sectors. Several developments lie behind this, including the growing importance of research and development, design, marketing, distribution and after-sales maintenance in the business processes of a wide variety of firms.

2. The growth in demand for such services has in part been fuelled by the need for knowledge and capabilities concerning the major new technologies that have been diffusing across the economy over the past few decades. New technology-related services – such as software, web services, biotechnology research and development, and materials engineering services, as well as a whole range of consultancy and training support services – are among the outstanding examples. Services play an important role in the innovation systems surrounding new technologies, as innovation researchers have belatedly come to realize. What has attracted little attention, though, is the way in which many of these service firms are internationalized; for example, some of the highest levels of employment in foreign-owned firms in Europe are in such services. This suggests that such firms play an important role in the international transfer of technology.

3. What has been more widely noted by innovation researchers is that some of these technologies have been particularly important to services, especially new information and communication technologies (ICT), which have been widely adopted in many services branches. There has also been interest, usually from other quarters, in the possibility that the technologies could facilitate internationalization of services. The trade 'enlarging' impact of new ICT in services has received much attention. The notion is that computer-communication systems in general, and the Internet in particular, can be used to enhance trade in information-related services and in the informational components of other services. These systems can also support internationalization by, for example, enabling closer management of overseas operations and franchises, and by allowing better networking among members of partnerships and expert-based professional services. As discussed in later chapters of this book, it may be that the intangibility of services has meant that much of the new trade enabled by ICT is in the form of intra-firm rather than arm's-length transactions. This underlines the relative importance of multinationals in services, and

also suggests the need for a critical examination of trade and foreign investment statistics.

4.  The liberalization of domestic and international policies around trade and investment in services has long been on the agenda, and received a substantial boost with the end of the Cold War and the decline of political faith in 'national paths' and national champions. Indeed, it was in sectors such as telecommunications – major users of new technologies – that we saw early pressures for the abandonment of ideas of natural monopolies and the necessity for vital services to be run by state enterprises. The 'deregulation' and opening up of markets for many other classes of services and utilities followed apace, not least because of pressure from multinationals (especially US financial services, which persuaded their government, the most powerful actor on the international scene, to push for liberalization of services trade). The European Union has been eager to support these developments, partly because of its own perceived comparative advantages, partly because of the belief that the dynamism of the European economy was being hindered by barriers to services trade (and by continuing sluggishness in exposing telecommunications and utilities to competition).

Despite the magnitude and rapid pace of these developments, there has been little scholarly analysis of the dynamics and implications of services internationalization. The innovation research community, in particular, has remained surprisingly disconnected, in the main, from those with expertise on services trade issues. Indeed, despite the awakening of interest in services and their role in innovation and internationalization, there is still a great deal of ground to be made up as against more established areas. Both theoretical and policy-related assessments of the service sector continue to neglect the importance of such issues as the impact of organizational and technological developments on the changing nature of the service sector, the increasingly dominant role of multinationals in the process of internationalization of services, and the impacts that these developments may have on services innovation and the role of services in the more general innovation processes.

## THE ROLE OF THIS BOOK

The objective of this book is to encourage and contribute to debate on a topic that has been largely neglected in the literature. It is a topic where available statistics are patchy and problematic, so that broad descriptive exercises are valuable (and raise important definitional and conceptual questions). It is also one where economic and social theory needs to be better developed to handle the new technological and organizational context for services and their internationalization, and the broad range of policy issues (not just trade policy) and strategic considerations (not just locational and outsourcing strategies) that follow from this. The essays in this book thus bring together the contributions of experts in internationalization of services and innovation in services. These experts are drawn from a number of disciplines – they include economists, sociologists with an interest in economics, international business specialists and geographers. Together they provide data analyses and conceptual frameworks that advance the discussion considerably.

The book investigates the following general questions:

1. What is services internationalization, and what are its drivers? What are the roles of technological innovation, changing policies and competitive contexts in facilitating, retarding or shaping this process?
2. What role do international transactions and foreign direct investment in services play in encouraging or hindering the development and diffusion of innovations (technological and organizational) within and between countries (and regions)?
3. To what extent do international transactions and foreign direct investment in services diffuse 'best practice' in technology innovation and management (and also in organizational innovation)?
4. To what extent do international transactions and foreign direct investment in services reinforce virtuous (or vicious) cycles of economic growth (or stagnation)? Can international transactions and foreign direct investment in services be a source of growth and structural adjustment to replace declining manufacturing industries? Are they necessary for a healthy manufacturing sector in the twenty-first century?

To address such questions, the chapters in this book adopt a variety of approaches, including the following:

1. appraisal of trade statistics – clarifying the many challenging features that emerge in terms of trends and national differences;
2. industry studies of knowledge- or technology-intensive service sectors which have experienced the greatest increase in international transactions and foreign direct investment – studies considering, for example, the magnitude of these processes, the changing industry structure and the various geographical regions participating in this process;
3. studies of multinationals which invest in services abroad – addressing

themes such as the technological infrastructure and the organizational forms they utilize in managing global services;
4. empirical and theoretical contributions concerning the effects of international transactions and foreign direct investment on economic competitiveness – examining their interaction with industrial activities and other services and influences on innovation processes and the more general context of services activities;
5. analysis of the policy implications of the internationalization of services in developed and developing countries – where issues of development and competitiveness are naturally raised.

## REST OF THE BOOK

The opening two chapters in Part I consider the theoretical traditions that have developed around the analysis of services innovation and internationalization. Miozzo and Miles argue that, despite parallel developments in analysis of internationalization of services and of innovation in services, and despite the evident relations between the two topics, these two strands of literature have rarely been brought together in any systematic way. The main approaches to international production in services fail to integrate notions developed from the economics of innovation. New technology is seen as an enabler, even a driver of internationalization, but the analysis is left there. Moreover, few contributions dealing with innovation in services grapple with developments in the internationalization of service activities, other than as an exogenous impact on the competitive environment faced by firms. Miozzo and Miles argue that the two strands of literature need to be brought together to address the current most pressing questions about internationalization, privatization and liberalization of services. In particular, this would enable the examination of the impact of international production in services on the innovative capacity of host countries and the ways in which a stronger services sector might assist a country's competitiveness in the supply of industrial products. The authors suggest elements of a research agenda that are required to foster an integration of the two approaches.

An equally important theoretical problem is identified in the chapter by Ietto-Gillies, again with implications for the analysis of innovation and internationalization. Ietto-Gillies discusses tangibility and productivity as demarcation criteria for services and manufacturing. New technologies challenge these demarcations by enhancing the complementarity between manufacturing and services, leading to organizational changes in all sectors of the economy and increasing the scope for internationalization. A new mode of internationalization is fostered – what Ietto-Gillies refers to as an 'electronically transmittable mode' which eliminates or at least reduces traditional space and time constraints. These developments may be resulting in a new international division of labour, with incentives for splitting the production process of many services into discrete components, some of which can be 'downloadable' and processed by pockets of skilled specialists in less developed or intermediate countries. Ietto-Gillies compares the internationalization of the world's largest transnational corporations (TNCs) operating in services and manufacturing. Two indices are presented and a data analysis suggests that services tend to be less internationalized than manufactured products. Ietto-Gillies describes the limitations of these types of analysis in the face of technological change. She calls for further research on the sources of productivity growth in the economy, on statistics to take into account the technology intensity of products and processes, and on the various forms of (international) delivery, including electronic ones.

These themes are picked up in Part II. In particular, the two chapters set out the limitations of available measures of international production in services, and also recognize the various modes of international services supply. The aim of Part II is to present an appraisal of available trade and investment statistics. These form fundamental backdrops to the later chapters of the book. The analyses in each chapter confirm many of the general statements regarding the intensification of internationalization in services as new technologies and policies are introduced. Baker (with Miozzo and Miles) examines the trends and patterns in trade and foreign direct investment of European services. Conventional trade in services (cross-border supply and consumption abroad) is important in many services, but it also significantly underestimates the importance of international production and transactions in services. Baker discusses the possible relationship between trade and foreign direct investment in services. In this respect, assumptions derived from the analysis of manufacturing trade and investment may not generalize well to many services. The prospects of greater services internationalization have so far only partly been realized, at least as documented by the available trade statistics. While services dominate the economies of the EU, trade in services in the EU accounts for only about 20 per cent of total trade. Investment, however, is another story. FDI in services accounts for about 55 per cent of total foreign direct investment flows and stocks: a figure much more commensurate with the importance of private service sector activities in domestic economies. Baker goes on to consider cross-national differences, finding that trade is more important for smaller member states, who have the highest trade to GDP ratios. This suggests that there might be stronger motivations in these countries for firms to trade (and undertake foreign direct investment) and for governments to create incentives to encourage trade and foreign direct investment in services.

Zimny and Mallampally suggest, on the basis of available time series data for the United States, that the internationalization of services as a whole and of a majority of individual service industries now takes place more through FDI than through trade. The relative importance of FDI as a mode of international delivery of services has increased in recent years. The proportion of local sales in total sales by foreign affiliates has risen noticeably for US TNCs operating abroad. This suggests that technological developments affecting services production and delivery have been more to strengthen the ability of firms to compete internationally through the establishment of operations in foreign markets than to strengthen their ability to compete through trade. However, within the category of services that can be delivered by both modes of delivery, trade has increased rapidly, along with FDI-related sales, pointing to increased complementarity of trade and FDI. Also available data suggest that TNCs in the USA are using their international production networks to exploit cost differences in services production across countries to improve competitiveness, but the scope of integrated production (as revealed by intra-firm transactions and affiliate-to-affiliate trade) is as yet limited.

As argued in Part I, new technologies may also be instruments for creating a new international division of labour around the production of services firms. This issue is explored in the contributions to Part III of the book. The capacity of TNCs to segment the production process of services and locate activities in different parts of the world may lead to new corporate and international division of labour. On the one hand, the standardization of certain service components may lead to certain segments of the service production chain being located where appropriate resources (mainly labour) are available and relatively cheap. On the other, there may be different organizational structures at the regional level, with some locations attracting the more high skill and strategic coordination and control functions. Understanding exactly how these trends are combined and intertwined presents huge challenges for analysis. Coe discusses the spatial complexity of TNCs' organizational structures in services, drawing on interviews with USA-based IT firms with bases in Singapore. Coe concludes that the processes of international expansion are complex in services. Notably, service activities in the firms presented are coordinated and organized at an intermediate 'regional' level between the 'national' and the 'global'. As TNCs increasingly organize their operations at the regional level, regional headquarters (places that attract the coordination and control functions) may become necessary structures to implement global strategies at the regional level. This is usually seen as a highly desirable form of foreign direct investment for host countries, owing to the relatively high quality direct employment they provide, the high levels of interactions with financial and business services in the host country, and the influence that may be brought to bear on investment decisions of the parent firm. Coe finds heterogeneous and complex organizational structures at the regional level. For example, Singapore, through being home to both the Southeast Asian or Asia–Pacific headquarters and support functions of many US IT firms, has attracted significant numbers of high-quality jobs (for regional sales, marketing and implementation teams). Overall, the US IT industry's investment in Southeast Asia reveals different dimensions of spatial variation occurring within a broadly multiregional structure – interregional variations, intraregional variations and intra-activity variations.

The new developments in the wider international division of labour are associated with a concentration of knowledge- or technology-intensive services in certain important cities, despite the claims that new technology frees us from the constraints of spatial location. Rubalcaba-Bermejo and Gago-Saldaña test the role of international and national effects in regional concentration of European business services. Many innovative services were traditionally located only according to national, regional or urban patterns. However, internationalization may be rendering international effects increasingly important in shaping the pattern of concentration of business services. The chapter presents comparative European results on the location of business services, relating them to key explanatory factors (qualifications, innovative performance, density and economic development). Eurostat data for five countries suggest that both national and international effects are relevant in determining the regional concentration of business services. The density variable and the number of patents per 1000 inhabitants turn out to be outstanding locational factors, when both national and international effects are controlled for. Business services location is closely linked to urban density, national profiles and regulatory factors, and, at the same time, to the degree of exposure to international competition. Also the correlation between business services and GDP per inhabitant is positively biased by the presence of international cities, which are the ones most exposed to international competition. The study provides a valuable starting point for considering the future development in space – and thus the internationalization – of business services.

The chapters in Part IV address this issue more directly with firm-level data. The chapters here contribute empirically to the analysis of trends, characteristics and the factors that enable and hinder the internationalization of services. The survey data, drawn from different countries, regions and sectors, enable a greater understanding of the strategy and modes of internationalization among services firms, as well as the location of their activities abroad. Roberts reconsiders the findings of her survey of internationalization of UK business services carried out in the early 1990s, in the light of recent technological developments that have facilitated the growth of resource-oriented internationalization within the service sector. The earlier study highlighted the market-oriented nature of the internationalization of business services. Since resource-oriented service activity is characterized by high levels of standardization, this would seem to be of limited value when customized and knowledge-intensive business services are being produced. Nevertheless, Roberts points out that, while core business services activities remain market-oriented, there is scope for resource-oriented activity in the production of highly standardized service components. International production can also be seen to be emerging, based on the location of skills in the form of centres of excellence, for example.

Henten and Vad use data from a survey commissioned by the Danish Ministry of Industry to show that, contrary to theoretically based expectations, exports are the predominant mode of internationalization by service firms in Denmark. This result questions the sharp division between manufacturing and services sectors. The surveyed firms unexpectedly regard their services as being relatively standardized, which gives us a clue to their exportability. Another result is that, to a higher degree than in other services, information-intensive services use telecommunications to deliver their services. As for barriers to internationalization, a major factor is not external constraints such as external conditions and regulations in foreign countries, but, rather, a lack of ambition to expand internationally which predominated in the vast majority of firms with no international sales. While policy makers and managers have mainly concentrated on the external obstacles to the international expansion of service firms, these results suggest that they should also consider the internal resources and motivations of firms. Information and communication technologies may also form important tools in the process of internationalization, which could also be foci of policy.

Toivonen's analysis of the internationalization of knowledge-intensive business services in Finland is based on extensive interviews with firms and professional associations. She shows the multiplicity of international activities and the new ways of operating in international markets. The study confirms that the degree of internationalization of knowledge-intensive business services is higher than in other services sectors. In some technology-based knowledge-intensive business services (such as software firms), the share of exports is important and growing, but Toivonen also points out the increasing importance of mergers and acquisitions as forms of internationalization. These results also underline the importance of chains of foreign origin, which hold a strong position in several knowledge intensive business sectors, such as advertising and auditing. These forms of internationalization, and even small-scale internationalization (that is, serving foreign clients in the home country) require strategic change, especially for small knowledge-intensive business services. Toivonen argues that the greatest difficulties for knowledge-intensive business services in internationalization are related to business skills. The results of the interviews call for an examination of the more indirect and non-equity forms of internationalization, such as assignments involving an international dimension, and international networks; these are particularly important in the non-technological knowledge-intensive business services.

Randles and Tether investigate the relationship between knowledge-intensive services and manufacturing in firms active in environmental technologies and services in the North West of England. Their findings show that services firms help manufacturers in the North West of England to gain access to overseas markets, but also help foreign manufacturers gain access to the UK and the North West of England regional market. In so doing, they play both a positive role in supporting 'cluster' development and a negative role in undermining this development. The chapter examines critically the benefits and drawbacks of a 'cluster perspective' for analysis and policy and suggests a critique of the 'stages' approach to the internationalization of knowledge-intensive services.

The chapters in this book provide a rich and diverse set of materials that contribute to the understanding of internationalization, technology and services. Through a combination of insights from different disciplines and methodologies, the contributions cast light on issues that are important parts of the general processes of globalization of production and innovation. In many ways, this marks the opening-up of a field of study. Both the implications of internationalization of services for innovation and the implications of innovation for the internationalization of services, need to be examined in more detail. We will need to get away from simplistic assumptions about the inevitable trade-enabling effects of ICT, or the boost to innovation that is yielded by international competition, and examine under just what conditions and how these tendencies are, or are not, manifest. Future research work needs to address such issues as the vulnerability of service operations to relocation through new technologies, the particular modes of location of service multinationals (and how these evolve) and the difference in strategy in different regions and sectors. The extent to which international services firms, especially knowledge- or technology-intensive services firms, add knowledge to (and enhance the productivity of) the alliances, clusters, regions and countries in which they are (or choose to be) located needs to be addressed further. Such research should help inform the answers to a final set of questions, which, perhaps more than this, depend on the scope and capability of national and pan-national governments to exercise political choice. What types of policy instruments can support a positive interrelationship between internation-alization and innovation in services? How can such instruments be gener-ated and implemented? What are their complementarities and contradictions with other policy goals, such as social equity and sustain-ability?

PART I

Innovation and internationalization of
services: conceptual issues

# 1. The relation between the internationalization of services and the process of innovation: a research agenda

**Marcela Miozzo and Ian Miles**

## INTRODUCTION

Services may remain elusive in many ways, but their importance has gradually impressed itself on researchers and practitioners from a wide range of backgrounds. Issues related to both the internationalization and the role of innovation in services have accordingly attracted increasing attention from social scientists, business analysts and policy makers.

Services internationalization has been the focus of much interest, in particular since the mid-1980s. At this time, the inclusion of services in the Uruguay Round of Multilateral Trade Negotiations began to stimulate statistical and analytic attention to the themes of foreign direct investment (and trade) and multinationals in services. A literature on international production in services has emerged. This draws attention to the distinctive characteristics influencing international production in services as well as to similarities with those of manufacturing. It addresses issues relating to national and international policy frameworks with respect to foreign direct investment and multinationals in services (see, for example, Enderwick, 1989, and the collection of articles in Sauvant and Mallampally, 1993).

Services innovation has also been an emerging area of attention over the same period. This literature argues that services should not be regarded solely as passive recipients of innovation from the manufacturing industry. Whether owing to information technology having triggered innovation across the whole economy, or to other developments, services are sources of innovation in their own right, as well as being in the vanguard of information technology use. This literature challenges the theories of 'post-industrial' society (see, for example, Gershuny and Miles, 1983). It stresses the peculiarities of services as technology producers and suggests policy recommendations relating to the scope of innovation in services.

Sometimes it suggests that traditional demarcations between services and manufacturing are irrelevant to innovation processes. Sometimes it suggests that the historical demarcations are eroding, for example, as a result of services becoming more technology-intensive, manufacturing incorporating new production practices, or the growth in producer services and the externalization of service functions by firms in other sectors (Miles, 1993).

Despite the parallel developments in documentation and analysis of internationalization and innovation in services, and despite the evident relations between the two topics, these two strands of literature have rarely been brought together in any systematic way. The main approaches to international production in services fail to integrate notions developed from the economics of innovation. New technology is seen as an enabler, even a driver of internationalization, but the analysis is left there. Similarly, few contributions dealing with innovation in services grapple with developments on the internationalization of service activities, other than as an exogenous impact on the competitive environment faced by firms.

This chapter will argue that the two strands of literature need to be brought together. This is essential if we are to address the most pressing questions regarding internationalization, technology and services, in the face of changes in the global economy. With this in mind, the first section outlines the contributions and the limitations of the application of theories of international production to service firms. The next section examines the contributions and limitations of the literature on innovation in services. The third section discusses the changing policy environment and its implications for the internationalization of services. The final section suggests elements of a research agenda that are required to achieve and build on an integration of the two approaches.

## INTERNATIONAL PRODUCTION AND SERVICES

Since the late 1980s, increasing effort has been made to apply theories of international production, which were originally based upon data collected in the manufacturing sector and service sector industries (Dunning, 1989; Enderwick, 1989). This literature was of particular interest to management researchers, though it also received contributions from several other sources (notably geographers). Many of the contributors identified important differences between the strategies and structure of services and manufacturing multinationals. While leading manufacturing multinationals have restructured their operations with an emphasis on 'downsizing', a renewed focus on core businesses and a growing dissatisfaction with conglomerate operations, the service sector has been characterized by continuing internationalization and increasing firm size and scope (Enderwick, 1992). This suggests that the challenges facing the expansion and growth of service multinationals may be quite different from those facing manufacturing multinationals (Campell and Verbeke, 1994).

One of the aims of this strand of literature has been to examine the firm-specific advantages of services multinationals. Competitive advantages, in the form of technology assets or the knowledge component of production techniques, are more difficult to identify in the case of services multinationals, because of the traditional concepts of technology and the traditional indicators of technology intensity such as R&D expenditures. The key intangible assets of services firms include the capability to acquire, process and analyse data, and the capability to coordinate skills, talent and expertise and their use across national boundaries (Sauvant and Mallampally, 1993; Enderwick, 1989; for a review of evidence on intangible assets and services, see Miles and Tomlinson, 2000).

The internationalization strategies of services firms have also been examined. For many services industries, particularly those that are mainly intermediate inputs into the production chain and those that are circulation activities, the initial stimulus to their internationalization was the rapid growth and global spread of multinationals in manufacturing sectors. The very early internationalization of services such as insurance, banking and accountancy, for example, may have been driven by the need to accompany the multinational operation of their manufacturing clients. As pointed out in Baker *et al.* (2000), it is not surprising that international operations of UK, Dutch and Swiss firms in such sectors are high because they are also home to a large number of multinationals in other sectors. More recently, however, the internationalization of manufacturing and services has become mutually reinforcing and the existence of an extensive global network of the well-known leading producer service firms in a country is seen as an important factor shaping the future evolution of multinational manufacturing activities (Roberts, 1999; Dicken, 1992).

This literature makes an important contribution in the way it draws a contrast between manufacturing and service multinationals in terms of their structure, internationalization strategies (including the degree of multinationality, risk, growth strategies, size and specialization) and derives different recommendations for best practice strategic management for multinationals in services. It has paid less attention, however, to the relations between the form and patterns of international production and trade in services and the impacts of these developments on different regions. One partial exception is the work carried out by consultancies, such as Dorothy Riddle and colleagues in Services Growth Consultants, who have also produced several guides for the United Nations Conference on Trade and

Development (UNCTAD). Such studies highlight the prospects for specific countries, islands and regions. Another good example is the set of studies prepared by Shrimpton and colleagues for Newfoundland (Shrimpton and Pollett, 2000; for a full report, see *http://www.mun.ca/cibs/EKBS/ekbs.htm*).

Another important exception includes the research on the internationalization of services that has been carried out mainly by geographers (see, for example, Howells, 1990, regarding the internationalization of R&D; Perry, 1990, on advertising; Warf, 1995, on telecommunications; Dicken, 1992, on financial services) and on the internationalization of services multinationals (see Clairmonte and Cavanagh, 1984) as a result of developments in information technology and in changes in government regulation. Changes in information technology carry important implications for multinationals. Economies of scale and scope enabled by new technogies are accompanied by a dramatic increase in the centralization capabilities of parent firms over their worldwide activities. This is expected to reinforce particular patterns of international division of labour at the corporate and regional level (Martinelli, 1992a; Vaitsos, 1988; Miozzo and Soete, 2001). This may include having peripheral branches only involved in data processing, dependent on central headquarters for knowledge and information transfer.

On the one hand, information technology facilitates the externalization of information processing and analytical operations to other, often smaller, firms (especially smaller firms that spun off from the parent firm that is now the client). On the other hand, multinationals may be able to reap the advantages of economies of scale and scope of service activities. Whereas technological standardization has been achieved by multinationals in terms of the physical infrastructure, and the international language of software and information technology skills, it is the institutional standardization that is now seen to be the main barrier for their further efficiency improvements (Miozzo and Soete, 2001). For such international service firms, it is seen as essential not to be placed at a disadvantage vis-à-vis local suppliers. They consequently pursue, in the first instance, the rights of establishment, equal national treatment and free access to information from their central databases through multilateral trade negotiations (ibid.).

Assessment of these issues raises the broad question of how increasing internationalization of services and the operation of multinationals in this area interact with the process of creation and diffusion of innovations in different countries or regions – including the role of information technology. In general, however, research on the internationalization of services has evolved independently from the debates over 'national systems of innovation' (Freeman, 1987; Lundvall, 1992; Nelson, 1993) and the globalization of production and innovation. As a result, little attention has been paid to the effect of the growing internationalization of services firms on the national systems of production and innovation of different countries. (For a general discussion of the relation between multinational firms' strategies and national systems of innovation, see Chesnais, 1992.) It is therefore difficult to respond to pressing policy questions, with empirically based results, as opposed to theoretical assertions. For example, what role do international transactions and foreign direct investment in services play in encouraging and hindering the development and diffusion of innovation within and between countries (and regions)? What do they mean for the ability to generate local business and employment and contribute to local knowledge creation?

More general studies suggest that multinationals are rather reluctant to locate technological activities in host countries (Pavitt and Patel, 1999). When firms move their innovation activities abroad it is to exploit technological advantages of host countries (Cantwell, 1995; Dunning and Wymbs, 1999). However, these issues regarding globalization in the production and commercialization of knowledge have not been investigated in relation to the internationalization of service activities. Moreover, a broader political economy approach is absent. Arguably, it is important to study the distribution of costs and benefits of the globalization of services within and across national boundaries. In particular, public policy needs to be underpinned by an approach that is able to distinguish investment in a country designed to create new technological capabilities from that allocated solely for mergers and acquisitions. Also the global generation of innovations by multinationals may aggravate imbalances between countries and regions – the link between the global and the local may need to be supported by public intervention. (In practice, what often happens is an unseemly competition between cities, regions or countries as to which can provide the best inducements to major foreign firms proposing to invest in their part of the world.) By not examining these issues, there is a danger that governments may view their role as simply providing 'externalities' to multinationals, creating environments conducive to business. Governments might then be abdicating any role in moderating or reshaping the very strong processes working towards increased differentiation (both social and across regions), unequal development and increasingly hierarchical domestic and international economic order (Chesnais, 1992).

## INNOVATION IN SERVICES

Services innovation is a topic of growing interest for innovation researchers and policy makers. The idea that services are labour-intensive activities,

with little scope for rapid productivity growth, and with little native innovative capacity, is clearly a notion of the past. Data of various kinds – case
studies, R&D statistics, innovation surveys – and the evidence of one's own
eyes in using banks, telephone services, accountants and so on have made
services innovation inescapable.

Research into services innovation can be grouped into three approaches
(Coombs and Miles, 2000). The first approach draws a demarcation
between services and manufacturing and stresses the peculiar features of
services, such as intangibility, interactivity and information intensity. This
approach often leads proponents to call for specialized studies of services
and for the development of novel instruments and theories for services
innovation. The second approach, assimilation, suggests that services innovation is fundamentally similar to its equivalent for manufacturing firms
and products. It can thus be studied effectively using the methods and concepts developed for manufacturing. This approach requires only minor
modifications to conventional statistics and surveys and demands little or
no new conceptual development. Coombs and Miles argue that a third
approach, a synthesis approach, is most desirable, since examining services
is important not only for an appreciation of its distinctive features, but also
because it may illuminate neglected aspects of innovation processes which,
though displayed prominently by service firms and industries, often apply
to many sectors of the economy.

For a long time researchers into services innovation were strongly demarcationist, stressing the peculiarities of the process. The 'reverse product
cycle' approach initiated by Barras (1984, 1986a, 1986b) suggests that the
absorption of new information technology into services as a means of
increasing service process efficiency provides a catalyst for services to
undertake their own innovation trajectories. Thus from process innovations
they move through a period of quality improvement to product innovations. The pattern he describes is often equally well characterized as an evolution from back-office process innovation, through innovation in the
delivery of services to clients, to product innovation. Miles (1995) argued
that delivery innovation is an exceptionally important feature of current
services innovation, especially but not solely through the application of
information technology. But relations between service suppliers and their
clients include more than just the delivery of the service itself.

Another approach to services innovation stresses these relations, using
the term 'servuction' (Belleflamme *et al.*, 1986) to describe the relations that
surround the production process, product and indeed consumption. The
authors contrast innovation in servuction with innovation in production:
these are distinct sites for innovation, and may have specific processes and
trajectories. An implication is that many innovative activities involving

interactions between suppliers and clients (and networks of clients), such as marketing, input of information from clients for choice, design or customization of the service, and after-sales support, may be overlooked, underemphasized or misrepresented by conventional measurement approaches.

Finally, intellectual property protection is an issue in services innovation, especially since 'intangible' products may be hard to protect within systems that evolved historically to protect technological innovations. While some commentators argue that the weak intellectual property regime in services has acted as a serious deterrent to innovation, the empirical case is not clear-cut, and there are intense arguments about the impacts of protecting software, business processes and other service activities.

More recently, assimilationist perspectives have become more common and some authors suggest that innovation surveys demonstrate little difference between manufacturing and service firms. For example, the distinction between product and process innovation is meaningful in the case of services, despite the claims to the contrary of many service sector researchers (Sirilli and Evangelista, 1998). Admittedly, these surveys do not address many features of innovation where case study work suggests demarcation. For example, the organization of innovation in services seems to be quite distinctive (Sirilli and Evangelista, 1998), being less centred on R&D departments and managers, and more oriented towards project management, than has been found to be the case in manufacturing; but the surveys do not examine such organizational issues.

An attempt to align services and manufacturing innovation was made by Soete and Miozzo (1989). Pavitt (1984) had classified services as simply supplier-driven in his taxonomy of innovative firm and sector types. Soete and Miozzo, in contrast, show that several contemporary service branches readily fit across the range of a slightly modified set of the categories. This suggests that the range of innovative practices in services is as wide as that in manufacturing. Soete and Miozzo distinguish different categories of service firms: (1) supplier-dominated sectors: these include public or collective services (education, health care, administration) and personal services (food and drink, repair businesses, hairdressers and so on), together with the retail trade; (2) scale-intensive physical networks sectors: these involve large-scale back-office administrative tasks dependent on physical networks (for example, transport and travel services, and wholesale trade and distribution). Such services are especially suited to the application of information technology (IT) with the aim of (at least initially), reducing costs; (3) information network sectors: such sectors are dependent on elaborate information networks (for example, banks, insurance, broadcasting and telecommunication services). These services play a major role in defining and specifying innovations, influencing the suppliers of new technologies; (4) specialized technology suppliers and science-based sectors: these include such services as software and specialized business services, laboratory and design services. The main source of technology is the innovative activity of the services themselves.

This approach, then, moves towards being a synthesis: it demarcates types of service innovation, but within a framework of manufacturing innovation. Recent years have seen many more efforts to understand and classify the diversity of service activities (Evangelista, 2000; Hipp *et al.*, 2000; Silvestrou *et al.*, 1992; and see the review in Tether *et al.*, 2001).

This literature brings out aspects of innovation that tend to be neglected in studies of innovation in manufacturing, and particularly in the R&D management literature, that may become more prominent across the whole economy. Among the main findings is that services innovation is rarely organized in terms of 'standard' models of R&D management structures, and is typically conducted on a more ad hoc, project management basis. Moreover, service firms tend to be poorly integrated into innovation systems and make little use of innovation-related facilities offered by institutions such as universities, research institutes or government laboratories. (Miles, 1999, provides some UK evidence; subsequent studies with Community Innovation Survey data confirm this; see CRIC/IDSE/ISE, 2001.)

This literature derives policy recommendations to encourage innovation in services. Nevertheless, one of the problems with this literature and its policy implications is that many of the contributions on innovation in services have evolved independently from those on the internationalization of services and the globalization of economic activity. Because of the close relationship to productive activities, the growth of services – and producer services in particular – must be analysed and interpreted within the context of changes in the way production is carried out. As Martinelli (1992a, p. 2) puts it:

> In particular, the development of such activities must be considered the result of at least [three] major interrelated processes: (1) the progressive concentration of capital and the rise of the modern large corporation, multiproduct and multilocational; (2) the growing internationalization of markets and competition; and (3) the development of new information technology.

Caution must be exercised with respect to the sole encouragement of the growth of knowledge- or technology-intensive services. For example, Tomlinson (2001) demonstrates that the impact of knowledge-intensive services on overall GDP in the United Kingdom is relatively weak compared to the case of Japan, for example, even though the UK's knowledge-intensive business services appear more developed than in Japan and other countries. This evidence suggests that growth of these services in itself would not necessarily lead to major improvements in productivity. The nature of the economic system of which services are a part is of vital importance.

There have already been some contributions looking at different 'national systems', classifying countries in terms of their specialization in different service sectors, and these contributions begin to suggest ways of integrating approaches to services internationalization and innovation. One such attempt analyses employment and occupational structures and argues for two ideal models. On the one hand, the 'service economy model' (represented by the United States, the UK and Canada) is characterized by a rapid phasing out of manufacturing employment after 1970. This model emphasizes financial services over producer services, and has seen a dramatic rise in jobs in healthcare and, to a lesser extent, in education. On the other hand, the 'industrial production model' (represented by Japan and Germany) is characterized by a restructuring of manufacturing, with the bulk of service jobs in services to firms and in social services (Castells, 1996).

A second attempt (Landesmann and Petit, 1995) classifies countries in terms of trade in services. A first ideal type, the 'integrated' type (represented by Germany and Japan) corresponds to relatively closed economies with a low profile in service trade on most items, with imports outpacing exports by a large margin. Firms tend to produce in-house most of the services linked to productive activities. There may be extensive subcontracting but within branch organizations. Long-run contracts are favoured and cross-shareholdings may be taken, especially with regard to finance activities. Trade in (complementary) services is reduced as much as possible. The second ideal type (represented by the UK and France) corresponds to 'open' economies that resort, as much as possible, to external activities for all peripheral activity. There are imports of out-of-house services and specialization in international service transactions with much intra-service trade (in a similar manner to the Helpman and Krugman, 1985, model for manufacturing). Service firms of the country remain very much national in the sense that, even if they are present in all foreign markets, their operations remain basically national. In other words, the relative share of the activities of foreign affiliates remains relatively low. The profile for trade in services would display strong credits and debits. A third ideal type is marked by its contribution to 'globalization' (perhaps this is an idealized version of the USA). Many peripheral activities can be provided abroad, where they are needed, so external transactions tend to be reduced mostly for services that deal directly with the organization of production processes and of markets.

Some services of general concern, like communications, or those linked with assets and general long-run interests of the firm, would, on the contrary, be traded at a level remaining rather higher. Direct investment substitutes to some extent for trade in services since these can be bought up locally. However, some international trade flows reflecting these communications and expertise-based services still take place (Landesmann and Petit, 1995).

The services innovation literature needs to be more attentive to these 'national systems' approaches. This might enable us to move to a synthesis of the work on innovation and internationalization. One of the steps here would be to study the impact of internationalization and growth of especially high-technology services on the overall capacity of different economies to absorb innovations, preserving job quality and bringing about a subsequent increase in demand. To quote again from Martinelli (1992b, p.85), this time referring to producer services:

> To address them as independent activities, without taking their market into prior consideration, is a widespread mistake. It is not sufficient to state 'producer services must be developed', but one must ask, 'which services should be developed, for which industries?' This means that the industrial structure of the region must be carefully assessed and that any service policy must be strongly integrated with industrial policy.

So far there has been little research combining innovation in services with an understanding of different national systems or models. A notable step towards studying these issues is the work by Mason and Wagner (1998), who have undertaken international comparative work to examine the ways in which the outsourcing of specialist services (product development, software developments and systems integration) affect knowledge transfer in electronics production in Britain and Germany. This study is described by its authors as a preliminary analysis, and more attempts to examine such issues are a priority.

## CHANGES IN POLICIES AND SERVICES INTERNATIONALIZATION

One of the most important factors affecting the internationalization of services since the mid-1980s has been the revision of government policies towards services. Deregulation, liberalization and the impact of new information and communications technologies have resulted in the intensification of internationalization. These developments lie at the heart of the globalization of manufacturing and are increasingly relevant to the internationalization of services.

Deregulation in sectors such as financial services, air transport and tele-communications has promoted the international expansion of firms within these sectors and helped provide the necessary infrastructure for the internationalization of other industries. Regional intitiatives like the internal market of the EU, and new multilateral structures like North American Free Trade Agreement (NAFTA) and Southern Cone Common Market (Mercosur), have intensified the internationalization of services in certain regions. On a global scale, negotiations at the World Trade Organization (WTO) regarding the General Agreement on Trade in Services (GATS) aim to further liberalize markets and promote competition in services.

Domestic and foreign competition are increasingly regarded by policy makers as tools to enhance the efficiency and productivity of service industries, which in turn is seen as being important for general economic development. Openness to foreign direct investment is regarded as an important way to inject and maintain competition, especially in oligopolistic markets (Mallampally and Zimny, 2000). Liberalization with respect to foreign direct investment has also taken place in less developed countries, particularly those in Latin America, partly as a result of privatization programmes. This has led to impressive increases in foreign direct investment in services in countries such as Argentina, Brazil and Mexico. The completion of the Uruguay Round and the adoption of the GATS have provided a strong channel for further liberalization of developing countries' policies related to foreign direct investment in services, especially in telecommunications and financial services.

The Uruguay Round stimulated the writing of a large number of articles, position papers and briefing notes about services trade. The vast majority of these, represented, for instance, by pieces in journals like *The World Economy*, were by mainstream economists, and adopted an assimilationist perspective. Essentially, the argument was that services trade had not taken off for a variety of reasons, not least protectionism, and that the features of services as depicted by the (at this point fairly sparse) literature were nothing very remarkable, and did not pose any challenge to established analyses of trade liberalization. This dovetailed neatly with the pressures for including services in the General Agreement on Tariffs and Trade (GATT) that was forthcoming from international services firms (especially certain financial services) and their home countries (especially the USA and UK). The perspective continues to inform much current thinking, for example from industry bodies such as the Global Services Network and from international governmental organizations such as the World Bank, which has this to say on its interesting web pages on services trade (see *http://www1.worldbank.org/wbicp/trade/services.html*, accessed September 2000):

Although some interesting twists arise because of the way services are traded and regulated, the basic insights from the theory of trade in goods apply to trade in services. There are likely to be substantial gains from liberalizing trade in services, immediately and in the longer term.

In contrast to such reassurances, the intangible nature of many services and the need for a high degree of interactivity between clients and suppliers of services have led some authors to stress the differences between services and manufacturing trade, in terms we might refer to as 'demarcationist'. Early discussions of the Uruguay Round were a delicate business because of the close association between trade and foreign direct investment in services internationalization. Debate about the Multilateral Agreement on Investment (MAI) reached a fever pitch around the turn of the millennium, and is continuing to bubble away as we write. Services internationalization, as subsequent studies in this book make clear, is often accomplished through investment and acquisitions. The concern is that rules restricting national governments' abilities to set rules for investment could undermine social and environmental policies, for example, transferring more power over national affairs to multinationals.

Another major area of concern about services internationalization is the liberalization and privatization of public services. The application of models from international trade in manufacturing to services clearly hit greater barriers, not least from activists that defend public provision of services and believe that these may be more efficient and innovative in an environment sheltered from private sector competition. Activists have been highly critical of the WTO push to extend trade agreements to cover services (through the GATS), particularly on the grounds that vital public services would be undermined: there is a wide range of ethical and political positions here, generally seeking to demarcate some public services (especially health and education) from other commodities and activities. The controversy is liable to continue, and the intellectual case for these particular sorts of demarcationist argument is likely to be articulated further. (The World Development Movement is one organization beginning to elaborate such a case, in the context of concerns about the liberalization of public services and the impact on poorer countries; see *http://www.oneworld.org/wdm/campaign/GATS.launchevent.htm*).

The role of new information and communication technology in making it easier to separate the location and timing of (some) service production from its delivery and consumption has led to interesting discussions of the 'mode of presence' of service organizations in foreign markets (Vandermerwe and Chadwick, 1989). There is a continuum of modes of presence, ranging from foreign direct investment (FDI) to management control procedures exerted through third parties in the national economy, such as franchising, licensing and management agreements. Where services can be traded using electronic (or photographic, printed or optical) media, physical presence can in principle be limited, be mediated by the final delivery and distribution of the services by the retail or delivery of the service via telecommunications in particular, which loosens locational ties in the production and consumption of services by making some information components of services transmittable, as in the case of the use of automated telling machines (ATMs) in international cash card networks. IT can also be used in international management control systems, allowing for rapid appraisal and monitoring of local operations and circumstances through telematics networks, in maintenance activities (for example, remote computer diagnostics), and for 'offshore office services'. Daniels (1993) draws on the work of Vandermerwe and Chadwick to suggest that there are three main clusters of service activity. First, there are services exported through 'goods'. These are mainly services where low supplier–client interaction is required (though contact may be frequent, as in some on-line services). Often these are information services with high degrees of similarity to publishing. Second, there are services utilizing third party methods such as franchising, licensing and joint ventures. These require some investment and systems of management control, there may be more or less interaction with clients, and the functions include physical ones (fast food, transport) and information services providing control of computers and communications rather than simple access to data (electronic mail, some software). Third, there are the services mainly using FDI. Typically, these are services with high levels of supplier–client interaction. Internationalization requires high control over service production and delivery, and the associated goods. Many specialized business services fall into this category. Often such services begin to spread overseas in the wake of the multinational corporations that are their major long-term clients. They may or may not then begin to compete with domestic service firms in the host country.

Such perspectives have been developed by those particularly keen to examine the dynamics of services, with less attention to parallels in other sectors. Some of the issues about 'mode of presence' do clearly apply to many other sectors. Efforts to provide more of a synthesis of the issues surrounding services internationalization and more general processes of internationalization are relatively few. However, Nayyar (1989) may be a case in point. While accepting most of the classical arguments as to the benefits of trade, he also draws attention to unusual features of services. Noting the extreme heterogeneity of the service sector, he calls for more study to examine how services (and how different sorts of services) may fit into the general picture, taking into account standard principles such as transparency, reciprocity and sovereignty. It is open to debate as to how far the analysis has moved on since he wrote, though, as this book demonstrates, we do now have access to much better statistics, and an armoury of case studies.

## SOME ELEMENTS OF A RESEARCH AGENDA

International production and investment decisions of multinationals, their location of employment and training will all have important consequences for host countries, not least on the absorptive and innovative capacity of the economy. The importance of the service sector for economic growth and development has now been increasingly recognized, but little research has been done on the impact that FDI and multinationals in services can have on the innovative capacity of host countries and their role in economic growth and development. To what extent do international transactions and FDI in services reinforce virtuous and vicious cycles of economic growth? Can international transactions and FDI in services be a source of growth and structural adjustment to replace declining manufacturing industries?

The answer to these questions requires bringing together the two strands of literature described earlier. A combination of these approaches may help develop policy recommendations to support the integration of innovation in services and the role played by service multinationals in this process with the national systems of innovation. A great deal of the impact of multinationals in services is likely to be of an indirect nature, especially through forward and backward linkages. Producer services and knowledge-intensive business services are particularly important here. But, perhaps, even more important are the effects that multinationals can have on the competitiveness of domestic firms and especially on the efficiency of their production process. A key issue, therefore, is to examine to what extent service multinationals contribute to competitiveness and help to increase the efficiency of domestic organizations, be they in the services or industrial sectors, state-owned or privately owned.

If we do not integrate these strands of literature, few clear implications can be drawn regarding pressing debates about deregulation, privatization or liberalization. Ill-conceived measures (such as 'competitive deregulation' to attract multinationals) could have consequences that are no less detrimental than the continued application of outdated or inefficient regulatory controls. The challenge of assessing such issues implies studying, in different countries, the nature of the services industries and their cross-linkages with other sectors and their importance for the attainment of overall economic objectives. We suggest that the most fruitful approach to linking the two literatures is to build on 'synthesis' perspectives: accepting that we can learn important lessons from the features that services contribute to economic activity, trade, other modes of internationalization, technological and organizational innovation. But we should not assume that these features are restricted to services, or that they are necessarily always going to apply to them.

Services innovation policy needs to be integrated with general innovation and industrial policy. Also it requires an examination of ways in which a stronger services sector might assist a country's competitiveness in the supply of industrial products, with particular attention to the growing data content of services and policies to improve this. Concomitantly, the role of multinationals and the conditions under which such investment could be put to most productive use for a host economy would have to be reviewed. There is considerable literature on the benefits of trade (and foreign direct investment) but services tend not to be treated explicitly in these analyses. Possible benefits from attracting inward investment in services include technology and knowledge flows, direct implications for employment and production, positive spillovers in training of domestic workers, improvements in standards of management, development of linkages downstream with suppliers and upstream with distribution. However, negative consequences have received less attention. Particularly for less developed countries, an infant industry argument can be made to protect the development of indigenous technology-intensive service sectors (Miozzo and Soete, 2001).

More generally, questions arise as to whether inappropriate models and techniques may be being internationalized. Are best practices for services accommodating the needs of local production and innovation environments? Is the outcome a growing concentration of service supply among a number of multinational firms at the global level? Is this reducing the autonomy of national goverments vis-à-vis increasing bargaining power of multinationals? Is this increasing the capacity of multinationals to relocate or segment production across the world according to their strategic considerations and regardless of the local production and innovation needs? How far do technology strategies (for example, for standards) support the internationalization of services and the integration of service affiliates into national systems? Whatever the case, there are a host of questions for policy makers that have received inadequate attention from the research community. We hope that this chapter has helped in the posing of these questions.

# REFERENCES

Baker, P., N. Plaisier and V. Spanikova (2000), 'The Internationalization of European Services', background paper for the Competitiveness Report 2000, Netherlands Economic Institute, Rotterdam.

Barras, R. (1984), *Information Technology and Economic Perspectives: The Case of Office Based Services*, Paris: OECD.

Barras, R. (1986a), 'A comparison of embodied technical change in services and manufacturing industry', *Applied Economics*, **18**, 941–58.

Barras, R. (1986b), 'Towards a theory of innovation in services', *Research Policy*, **15** (4), 161–73.

Belleflamme, C., J. Houard and B. Michaux (1986), 'Innovation and research and development process analysis in service activities', occasional paper no. 116, EC, FAST, Brussels.

Campell, A. and A. Verbeke (1994), 'The globalization of service multinationals', *Long Range Planning*, **27** (2), 95–102.

Cantwell, J. (1995), 'The globalization of technology: what remains of the product cycle model?', *Cambridge Journal of Economics*, **19** (1), 155–74.

Castells, M. (1996), *The Rise of the Network Society*, Oxford: Blackwell.

Chesnais, F. (1992), 'National systems of innovation, foreign direct investment and the operations of multinational enterprises', in B-Å Lundvall (ed.), *National Systems of Innovation*, London: Pinter.

Clairmonte, F. and J. Cavanagh (1984), 'Transnational corporations and services: the final frontier', *Trade and Development: An UNCTAD Review*, **5**, 215–73.

Coombs, R. and I. Miles (2000), 'Innovation, measurement and services: the new problematique', in J.S. Metcalfe and I. Miles (eds), *Innovation Systems in the Service Economy*, Dordrecht: Kluwer.

CRIC/IDSE/ISE (2001), 'Analysis of CIS data on innovation in the service sector. Final report to European Commission DG12', CRIC, University of Manchester and UMIST, Manchester.

Daniels, P.W. (1993), *Service Industries in the World Economy*, Oxford: Blackwell.

Dicken, P. (1992), *Global Shift: The Internationalization of Economic Activity*, London: Paul Chapman.

Dunning, J. (1989), 'Multinational enterprises and the growth of services: some conceptual and theoretical issues', *The Service Industries Journal*, **9** (1), 5–39.

Dunning, J. and C. Wymbs (1999), 'The geographical sourcing of technology-based assets by multinational enterprises', in D. Archibugi, J. Howells and J. Michie (eds), *Innovation Policy in a Global Economy*, Cambridge: Cambridge University Press.

Enderwick, P. (1989), *Multinational Service Firms*, London: Routledge.

Enderwick, P. (1992), 'The scale and scope of service sector multinationals', in P.J. Buckley and M. Casson (eds), *Multinational Enterprises in the World Economy: Essays in Honour of John Dunning*, Aldershot, UK and Brookfield, US: Edward Elgar.

Evangelista, R. (2000), 'Sectoral patterns of innovation in services', *Economics of Innovation and New Technology*, **9**, 3.

Freeman, C. (1987), *Technology Policy and Economic Performance: Lessons from Japan*, London: Pinter.

Helpman, E. and P. Krugman (1985), *Market Structure and Foreign Trade*, Cambridge, MA: MIT Press.

Hipp, C., B. Tether and I. Miles (2000), 'The incidence and effects of innovation in services: evidence from Germany', *International Journal of Innovation Management*, **4** (4), 417–54.

Howells, J. (1990), 'The internationalization of R&D and the development of global research networks', *Regional Studies*, **26** (6), 495–512.

Landesmann, M. and P. Petit (1995), 'International trade in producer services: alternative explanations', *The Service Industries Journal*, **15** (2), 123–61.

Lundvall, B-Å. (ed.) (1992), *National Systems of Innovation*, London: Pinter.

Mallampally, P. and Z. Zimny (2000), 'Foreign direct investment in services: Trends and patterns', in Y. Aharonin and L. Nachum (eds), *Globalization of Services: Some Implications for Theory and Practice*, London: Routledge.

Martinelli, F. (1992a), 'A demand-orientated approach to understanding producer services', in P. Daniels and P. Moulaert (eds), *The Changing Geography of Advanced Producer Services: Theoretical and Empirical Perspectives*, London: Belhaven Press.

Martinelli, F. (1992b), 'Producer services' location and regional development', in P. Daniels and P. Moulaert (eds), *The Changing Geography of Advanced Producer Services: Theoretical and Empirical Perspectives*, London: Belhaven Press.

Mason, G. and K. Wagner (1998), 'High Level Skills and Knowledge Transfer in Britain and Germany: Electronics, Technical Consultancy and Systems Integration', paper presented at the Systems and Services Innovation Workshop, CRIC, Manchester, 17–18 March.

Miles, I. (1993), 'Services in the new industrial economy', *Futures*, **25** (6), 653–72.

Miles, I. (1995), 'Services Innovation: Statistical and Conceptual Issues', report to OECD NESTI Working Group on Innovation Surveys PREST, University of Manchester.

Miles, I. (1999), 'Services and foresight', *Service Industries Journal*, **19** (2), 1–27.

Miles, I. and M. Tomlinson (2000), 'Intangible assets and service sectors: The challenges of service industries', in P. Buigues, A. Jacquemin and J-F Marchipont (eds), *Competitiveness and the Value of Intangible Assets*, Aldershot, UK and Northampton, MA, USA: Edward Elgar.

Miozzo, M. and L. Soete (2001), 'Internationalization of services: A technological perspective', *Technological Forecasting and Social Change*, **67** (2), 159–83.

Nayyar, D. (1989), 'Towards a possible multilateral framework for trade in services', in UNCTAD (ed.), *Technology, Trade Policy and the Uruguay Round*, New York: UNCTAD.

Nelson, R. (ed.) (1993), *National Innovation Systems*, New York: Oxford University Press.

Pavitt, K. (1984), 'Sectoral patterns of technical change: towards a taxonomy and a theory', Research Policy 13, 343–73.

Pavitt, K. and P. Patel (1999), 'Global corporations and national systems of innovation: who dominates whom?', in D. Archibugi, J. Howells and J. Michie (eds), *Innovation Policy in a Global Economy*, Cambridge: Cambridge University Press.

Perry, M. (1990), 'The internationalization of advertising', *Geoforum*, **21** (1), 35–50.

Roberts, J. (1999), 'The internationalization of business service firms: A stages approach', *The Service Industries Journal*, **19** (4), 68–88.

Sauvant, K. and P. Mallampally (1993), *Transnational Corporations in Services*, United Nations Library on Transnational Corporations, United Nations, general editor John Dunning, London: Routledge.

Shrimpton, M. and C. Pollett (2000), 'Small places, big ideas: Exporting north Atlantic expertize', in G. Baldacchino and D. Milne (eds), *Lessons from the Political Economy of Small Islands: The Resourcefulness of Jurisdiction*, Basingstoke: Macmillan.

Silvestrou, R., L. Fitzgerald, R. Johnston and C. Grant (1992), 'Towards a classification of service processes', *International Journal of Service Industry Management*, **3** (3), 62–75.

Sirilli, G. and R. Evangelista (1998), 'Technological innovation in services and manufacturing: results from Italian survey', Research Policy 27, 881–99.

Soete, L. and M. Miozzo (1989), 'Trade and Development in Services: A Technological Perspective', MERIT Research Memorandum 89–031. MERIT, Maastricht, The Netherlands.

Sundbo, J. (1998), *The Organization of Innovation in Services*, Cheltenham, UK and Lyme, US: Edward Elgar.

Tether, B.S., C. Hipp and I. Miles (2001), 'Standardization and particularization in services: Evidence from Germany', *Research Policy*, **30** (7), 1115–38.

Tomlinson, M. (2001), 'A new role for business services in economic growth', in D. Archibugi and B-Å. Lundvall (eds), *The Globalizing Learning Economy*, Oxford: Oxford University Press.

Vaitsos, C. (1988), 'Transnational Rendering of Services, National Development and the Role of TNCs', UN.

Vandermerwe, S. and M. Chadwick (1989), 'The internationalization of services', *Services Industries Journal*, **9** (1), 79–93.

Warf, B. (1995), 'Telecommunications and the changing geographies of knowledge transmission in the late 20th century', *Urban Studies*, **32** (2), 361–78.

# 2. Internationalization and the demarcation between services and manufactures: a theoretical and empirical analysis

## Grazia Ietto-Gillies

## INTRODUCTION[1]

Most economic and social changes of the last two decades have their roots in two main developments: the increased degree of international integration worldwide, and the introduction of new technologies, particularly information and communication technologies (ICTs). These two developments are interconnected in that the globalization process could not have taken off without ICTs. Moreover, the rate of diffusion of the new technologies is greatly enhanced by the economic forces behind the globalization process and in particular by the activities of the transnational corporations (TNCs).[2]

These developments have profound effects on a variety of economic and social elements including the introduction of new products and processes, the need for new skills, the development of new forms of organization of production within firms and industries and of stronger linkages between products and between industrial sectors. This chapter is concerned with the latter effect and in particular with the analysis of the way internationalization and innovation interact with the demarcation between services and manufactures.

Demarcation and taxonomy in general are theory-driven and this is true also in the case of the demarcation between services and manufactures. In our case the issue is now further complicated by the impact that the new technologies are having on the nature of products and their relationship with each other, the production processes and the organization of production in time and space.

The chapter will briefly consider two main demarcation criteria based on the tangibility of the products and on the productivity of the sectors, in sections two and three respectively. It will examine issues arising from the demarcation criteria, particularly those related to innovation and to internationalization. Section four analyses the implications of these developments for the international division of labour. The chapter will then go on to consider different theoretical frameworks for the analysis of internationalization and will introduce two specific indices (sections five and six). This is followed by an empirical study which assesses the degree of internationalization of manufactures versus services as it emerges from the direct activities of the world's largest transnational companies (section seven). The last section will draw implications and conclusions.

## DEMARCATION CRITERIA AND THEIR IMPLICATIONS: TANGIBILITY

The demarcation criterion most used for manufactures and services is related to a characteristic of the product: its tangibility or materiality (Grubel, 1987; Enderwick, 1992; Clegg, 1993). There are several reasons why this characteristic has analytical and practical relevance, and in particular the following ones. Unlike material products, immaterial ones cannot be stored and thus their transferability in space and time is impaired. This has implications for the tradability of products and for firms' strategies on how to meet demand in space and time. It has also implications for the space and time relationship between producers and consumers or 'providers' and 'receivers' of services: the extent to which they have to be in the same place and at the same time and if so who is to move and where (Stern and Hoekman, 1987; Sapir and Winter, 1994; Roberts, 1998).

Not everyone is happy with the demarcation criterion based on tangibility. In particular, Hill, in a widely cited paper (1977), defines a service not in relation to its immateriality but in relation to what it does to the receiver, whether a person or a good. This definition is effect-centred rather than product-centred and focuses on the interactivity between producer and consumer. It therefore overcomes the problem of material versus immaterial component of a product. Hill (ibid., p. 318) writes:

> A service may be defined as a change in the condition of a person, or of a good belonging to some economic unit, which is brought about as a result of the activity of some other economic unit, with the prior agreement of the former person or economic unit.

Moreover, the characteristic of materiality does not necessarily provide such a strong clear-cut case for demarcating between manufactures and services, for various reasons. First, most immaterial products involve elements of material ones for their full delivery either as part of the production process or as part of the final product. The production process leading to a service always involves material products: buildings where the service takes place; scissors for the hairdresser to cut my hair. The final product is very likely to have both a good and a service component (Gray, 1990; Grubel,1987): food to be served in a restaurant; paper (book) to embody the result of an intellectual effort; computers for the financial analyst; discs to embody software or music. This means that it is not easy to disentangle the contribution of the material from the immaterial component in the final product. In other words, there is a strong complementarity between manufacturing and service products.

Moreover, in terms of organization of production, any firm which is classified as manufacturing (or service) will be heavily involved in services (or manufacturing) products as well.[3] Because all firms – no matter what the nature of their final product – need business services as well as material products such as offices, vehicles and computers.[4] Nonetheless, in service products we do have a full or partial element of intangibility and its implications for tradability in time and space (including tradability across frontiers) are huge: immaterial products cannot be stored and sold at a different time or in a different location. They therefore exhibit less flexibility in the mode of delivery and in the timing of supply to meet demand.

Traditionally, this characteristic had two major implications, one for technical progress and one for internationalization. On the first one, Bhagwati (1984) developed an analysis of the way in which technical progress has often led to the 'splintering' of goods from services. In his view, the 'progressive' part of the old service would be incorporated in a material product leaving behind a reduced and 'unprogressive' service. Material products embodying traditionally immaterial services have often been developed in order to achieve storability and flexibility of delivery in time and space; examples include records and record players.

As regards the organization of production across space, the immateriality of the product makes it impossible to store and transport it to other locations including other countries. Conventional exports and imports are not possible, given the nature of the product. This means that foreign demand must be met either by the consumers moving to the production location or by production being moved where the consumers are. The latter may involve direct production in another country by the original producers – foreign direct investment – or through an intermediary via licensing and franchising.

This picture sees services as the Cinderella in the innovation field: they are lagging behind manufactures and they are seen as unprogressive unless they can be incorporated into a material product. However, information and communication technologies are dramatically changing this picture, for a variety of reasons.

First, ICTs have increased the services content of production in all the sectors of the economy and are further blurring the distinction between manufacturing and services. Manufacturing products are becoming more service-intensive and service output is becoming more manufacturing intensive (Miles, 1993; Nayyar, 1988; Kitson and Michie, 1996) – in particular, because both goods and services need the services of computer specialists as well as the hardware and infrastructure required by ICTs. Thus manufacturing and services are acquiring a stronger complementarity characterized by new qualitative elements in which ICTs are increasing the scope for linkages between products as well as sectors. It also increases the scope for the diffusion of innovation between industries (Tomlinson, 2001).

Second, ICTs are giving scope for further splintering of goods from services (to use Bhagwati's terminology): information-intensive services can be embodied in a compact disc which can then be supplied to consumers spatially located elsewhere from the producer. Third, and most relevant, is the fact that ICTs are also creating a new mode of delivery for information-intensive products: an electronically transmittable mode which completely eliminates space and time constraints. These effects of innovation have considerable implications for internationalization and we will return to this issue in the fourth section.

## DEMARCATION CRITERIA AND THEIR IMPLICATIONS: PRODUCTIVITY

The other main demarcation criterion between manufacturing and services is based on the contribution that the two major sectors of the economy make to the performance of the economy via their contribution to employment, development and growth.

Traditionally, manufacturing was considered to have wider scope for technical progress which resulted in higher levels and growth rates of productivity in manufacturing (Verdoon, 1949; Kaldor, 1967) compared to agriculture and services.[5] Indeed, this was the reason for the current sectoral taxonomy as originally developed in the works of Fisher (1939) and Clark (1940). The sectoral structure of production developed by economists was linked to the employment potential and to the process of development. The traditional taxonomy therefore reflected the underlying theory behind the explanation of development and employment potential.

Is this traditional taxonomy still appropriate in the era of ICTs? Is the demarcation 'manufactures versus services' the best one to capture the contribution of industries and sectors to the economic performance of countries? Three major elements due to the ICTs are creating the need to rethink this traditional Fisher–Clark sectoral taxonomy. First, there is the wider scope for increased complementarity between manufacturing and services mentioned in the previous section. Second, there is the fact that the introduction of new technologies goes hand-in-hand with changes in the organization of the production process in all industrial sectors and that, indeed organizational as well as technological changes contribute to the performance of various industries.

The third element, and probably the most important one, is the actual and potential contribution that the widespread use of ICTs can make to productivity and growth. Before we consider this issue, a word on the problem of the ICTs and the productivity changes is in order.

There is a prima facie expectation that the widespread use of information and communication technologies (ICTs) would affect the productivity of both manufacturing and services. However, until recently there has been no conclusive evidence about the effects of ICTs on productivity. Some literature has pointed out that, even in countries with high ICT intensities, productivity does not appear to have increased substantially and/or throughout many industries. In 1987, Robert Solow, while discussing the slowdown in productivity worldwide, came out with his now famous paradox: 'You can see the computer age everywhere but in the productivity statistics' (p.36).

Scepticism is shown, more recently, also by Gordon (2000), who maintains that increase in productivity outside the production of computers is very limited and destined to remain so because the human factor – necessary for the analysis of problems and results – causes diminishing returns even in the face of exponentially increasing computer power.[6]

However, in spite of these sceptical positions, evidence of increased productivity is now beginning to come through at a fast pace. Tomlinson (2001), reviewing several empirical studies, reports a 'highly significant relationship between value added, gross output and productivity and the value of knowledge-intensive business services purchased by each sector after taking into account labour and capital' (pp. 102–3). He refers to knowledge-intensive business services (KIBS) rather than the ICTs. However, these services are very ICT-intensive.

There is evidence also from work specifically related to the ICTs.[7] In particular, Oliner and Sichel (2000) find that the contribution to output and productivity growth (in the US non-farm business sector) of computer production and of the use of information technology (which includes computer hardware, software and communication technology) was very low in the early 1990s. However, it appears to have become very substantial in the second half of the decade. Indeed, on the basis of their empirical results, the authors estimate that information technology accounted for about two-thirds of the step-up in labor productivity growth between the first and second halves of the decade (p.21). They predict a continuation of this performance in years to come, particularly as the productivity gains from e-commerce are also likely to come on stream. Oliner and Sichel's conclusion highlights the discrepancy between short- and long-run results. This has both an empirical and a theoretical basis.

Dalum *et al.* (1999) point out that the nature of ICTs requires profound and widespread changes in the economy and society. This means that the full impact, including large and widespread increases in productivity, requires a considerable degree of adaptation and developments in society. A longer time may therefore be needed for the full impact to be felt.

Brynjolfsson and Hitt (2000), in an empirical micro study, stress the relevance of organizational changes and training for the full positive impact of the ICTs on productivity to be felt. Their firm-level studied point to organizational complements such as new business processes, new skills and new organizational and industry structures as a major driver of the contribution of information technology (p.45). All this involves the firms in extra expenditure in the short run in order to implement the necessary organizational changes and training programmes as well as meet the cost of hardware. However, it brings large productivity gains in the long run. The authors also stress the possible lack of consistency between micro and macro data on productivity and the need to concentrate on micro studies to get the full picture. The increased productivity and growth potential are brought about by the ICTs for a variety of reasons, including reduction in costs per unit of output, increased investment opportunity, and facilitation of complementary innovation across firms and industries (Brynjolfsson and Hitt, 2000).[8]

Thus several authors stress the positive effects of ICTs on productivity. However, the introduction of organizational changes alongside ICTs seems to be a necessary condition for reaping the full productivity effects. In the pre-ICT era, organizational changes at the firm and industry levels were part and parcel of technological changes whether brought about by economies of scale (Bhagwati, 1984; Stigler, 1951) or by the desire to cut down market transaction costs (Coase, 1937; Williamson, 1975, 1981). In the new technological environment, organizational changes are seen to be as necessary as technological ones in order to gain increases in productivity. The combination of technological and organizational innovation takes time and therefore productivity improvements may have longish lags in relation to the introduction of the new technology. This explains why the growth in productivity appears to be significant only from the mid-1990s onwards.

The strong and increasing role of the new technologies in the production processes means that high levels and growth of productivity are increasingly more likely to be linked to the intensity of use of such technologies, whether they occur in the production of goods or services: productivity growth may no longer be the prerogative of manufacturing only,[9] therefore the scope for analyses of development and growth based on the Fisher–Clark sectoral demarcation can be called into question. A new demarcation based on technology intensity (Preissl, 1995) and usage that cuts across the manufacturing versus services divide may become more appropriate for the analysis of development, growth and employment potential in the twenty-first century.

## A 'NEW INTERNATIONAL DIVISION OF LABOUR' FOR THE ELECTRONIC AGE?

The effects of ICTs on productivity are bound to have considerable impact on the firms' competitive advantages as well as on countries' comparative advantages and thus on the pattern of international specialization. In the pre-ICT era, developing countries were 'specializing' in primary products though there were manufacturing pockets linked to low-skills and labour-intensive processes. The new technologies are bringing about a new phenomenon, albeit at a small level of development as yet: the location of some specific skills-intensive components in developing countries.

Section two considered the introduction of e-transmission as a way of trading information-intensive services. This is indeed also a completely new internationalization mode as the service can be transmitted electronically within and across borders. This new mode can be used to trade final products (for example, to deliver a final report or music) or it can be used to transmit component(s) of a service that can be further processed in a different location – within the same country or abroad – and returned electronically to base.

Why would a firm want to have part of the product processed in other location(s)? There are two main reasons: in order to use labour which is skilled though cheaper than in the main production location, and in order to gain access to specific highly-skilled labour which is unavailable in the main production location. Either of these two reasons applies within and across nation states. Many developing countries have pockets of skilled labour which can be bought at much cheaper rates than the corresponding ones in developed countries: they range from accountants to software engineers, to data and text processors, to copy editors.

This creates incentives for splitting the production process of many services into discrete components, some of which can be downloaded and processed by pockets of skilled specialists in developing or intermediate countries. The organization of production within and across countries will therefore be done in accordance with the skills requirement of components and the availability and costs of purchasing such skills on the labour markets of the world.

We see here an application of the principles of 'new international division of labour' (NIDL) put forward in the 1970s and early 1980s (Fröbel *et al.*, 1980). According to this theory, the interests and activities of TNCs led to new developments in the production process by which the product would be divided into several components according to the skills required. The components requiring low skills would be located in developing countries (often in export processing zones) and those requiring high skills would be located in developed countries. One consequence of this NIDL is the increase in international trade as components are moved from country to country. Such trade would often be intra-industry as well as intrafirm because the components belong to the same industrial category as the final product and are transferred internally to the firm though across frontiers.

A similar pattern is now developing for many services – or services components of goods – because of the new technologies. We could be talking of an electronic-age 'new international division of labour' (E-NIDL) in which the ICTs allow increased scope for the conditions which led to the NIDL. In the E-NIDL the scope for division of the manufactured products into appropriate components is still applicable. However, to this we must now add the scope for division of services into components some of which can be transmitted electronically for further processing in other locations. Thus the E-NIDL comprises two elements: a manufacturing element to which the NIDL applies and a service element to which the e-transmission mode applies.

There are many similarities and differences between the NIDL and the E-NIDL. Both originate through the activities of TNCs; both involve division of the production process into discrete components and an appropriate organization of production within and across countries; both involve activities in developed and developing countries; both have an impact on the international division of labour and on trade; both are likely to give rise to intrafirm and intra-industry trade.

However, there are also some major differences. The NIDL refers to manufactures while the E-NIDL applies to services or services components of manufactures. The scope for internationalization increases in the E-NIDL as new forms and modes (through e-transmission) are added in the E-NIDL to the traditional imports and exports of the NIDL. In the NIDL, the division of the production process into various components is done in such a way as to allow the utilization of cheap unskilled labour. In the E-NIDL, the strategic element in the design of the production process for services is the utilization of cheap skilled labour. Pockets of such labour can, in fact, be found in developing or intermediate countries (such as India or Ireland) at much lower rates than in developed countries.

Though stress is laid here on the international division of labour, some of the points made are more general and apply to the spatial division of labour. They therefore apply to regions – within nation states – as well as between nation states. Thus, for example, the new technologies allow the location of call centres in places within the nation state spatially separated from the headquarters of the business or its main production location(s).

Many of the developments mentioned in this section are still at an early stage. Nonetheless, we already see some clear signs now: pockets of specialization in relatively high skills are to be found in Singapore (financial services), India (software) and Brazil (engineering) (Miozzo and Soete, 2001). Scotland and the North of England seem to have the lead on UK call centres. The rapid diffusion of the new technologies and the coming on stream of the full effects of these developments are likely to increase their relevance in the years to come.

## THE INTERNATIONALIZATION OF SERVICES VERSUS MANUFACTURES

The introduction and widespread use of new technologies together with the existence of other favourable conditions (John *et al.*, 1997; Dicken, 1998) have led to a tremendous increase in internationalization for all the sectors of the economy and in all it aspects and modes. This by itself has produced a bandwagon effect on services, particularly from the 1980s onwards (Mallampally and Zimny, 2000).

So how internationalized is the service sector compared to manufacturing? Before we can begin to tackle this question we need to clarify two conceptual issues. First, there is the fact that, ideally, the assessment of the degree of internationalization should take account of the different ways of reaching foreign customers. These range from production abroad directly via foreign direct investment (FDI) or indirectly via franchising and other contractual arrangements; export of services embodied in material goods such as CDs; movements of the customers to the producer's country as in tourism; delivery via the temporary movement of expert labour; and electronic transmission.[10]

Second, we need to clarify the spatial/cross-countries dimension in the assessment of internationalization. At the micro level, this is usually conceptualized as the degree to which firms' activities are located abroad rather than at home. In this framework indicators of internationalization assess the *degree of foreign projection* of the firms' activities, whichever of the activities are considered and, indeed, whether there is a single activity and variable or several. The indicators are usually constructed as ratio of foreign to total (foreign and domestic) activity. This is the approach taken in most indicators of internationalization, whether seen as a single element/variable or as multiple elements/variables (Dunning and Pearce, 1985; Sullivan, 1994; UNCTAD, 1995). In this approach we get the same degree of internationalization (for the firm and/or activity) whether the foreign activity takes place in one, a few or many foreign countries.

A different way of looking at internationalization is to consider the geographical extent of operations specifically in terms of number of nation states in which production or sales or business activities in general take place. This conceptualization sees the *degree of geographical spread* of activities into the various foreign countries as a relevant element in assessing the degree of internationalization (Ietto-Gillies, 1998, 2001, ch.4; Palmer, 2001). The theoretical relevance of this approach stems from the key role played by the nation state(s) in the decisions of transnational corporations (TNCs).

There are several dimensions to producing in different countries. First, there is the purely geographical/spatial dimension and therefore the spatial distance between business points, whether they be plants or production sites in relation to markets or in relation to suppliers or distributors. This is a question of spatial distance and transport costs and it applies almost equally between and within frontiers. There is also a cultural dimension in which the distance tends to be greater between nation states, though this is not always the case. Third, there is a regulatory dimension linked to the fact that different nation states tend to have different 'regulatory regimes', by which is meant a set of specific regulations which apply to people, firms and institutions within the borders of the nation state. Some of these regulations stem from the legal or institutional system, some from government policies.

Different nation states are characterized by different regulatory regimes. Within each nation state the same – or a more uniform – set of regulatory regimes tends to apply. Do firms benefit from operating across different regulatory regimes, over and above any advantages they may have of operating in different geographical locations and thus over and above any advantages deriving from the efficient use of resources available in each location? There may indeed be extra benefits from operating across different regulatory regimes. These may derive from three sources: (1) enhanced scope for the manipulation of transfer prices and thus for taking advantages of different tax and currency regimes; (2) spreading of risks linked to the political situation in different countries; and (3) strong bargaining power towards governments and labour – in particular, taking advantage of different labour regimes, for example the fact that labour can organize itself more easily within nation states than across them.

A strategy of wide locational network of production diminishes the risks of disruptions through industrial action. Moreover, and most important, it also fragments the labour force employed by the same firm, as labour is, on the whole, unable to organize across different countries. Such a strategy may, therefore, diminish the bargaining power of labour compared to a situation in which all or most of the firm's production is located within one or a few countries (Cowling and Sugden, 1987; Peoples and Sugden, 2000; Ietto-Gillies, 1992, ch.14, 2000, 2001, ch.6). A wide network spread of direct activities may also give credibility to any threat of relocation of production. It can therefore be used also when bargaining with governments for special incentives designed to attract inward investment.

The difficulties faced by labour in organizing across nation states are in stark contrast to the transnational firms' position: the TNCs are able to plan, organize and control production activities across nation states: this is what the TNC is about.[11]

Any strengthening of the firm's bargaining power towards labour and/or governments is likely also to have positive effects on its power over rivals. However, there are also costs associated with a strategy of locational spread in terms of possible missed economies of scale as well as higher managerial and organizational costs (Hymer, 1976).

A strategy of spreading direct activities by the host country may also bring advantages of knowledge acquisition and diffusion.[12] This might be knowledge about markets and production conditions to be utilized for future direct investment plans or for alternative modes of organization of production.[13] Moreover, the knowledge acquisition may lead to innovation processes of both organizational and technological types. There are knowledge spillovers internal to the firm, within and across countries as well as within the industry in the countries where the TNCs operate (Cantwell, 1989, 1995).

Is there an a priori link between internationalization mode and the cross-countries dimension of internationalization just discussed? There are some. Whether we accept a relationship of substitution or complementarity between trade and FDI (Cantwell, 1994; Ietto-Gillies, 2001, ch.2), the cross-countries network extensity may tend to be the same for exports and FDI. Nonetheless, in the case of substitution in the relationship, there is a linear time sequence and therefore the extensity is not contemporaneous, while in the case of complementarity it is.

The e-transmission mode for information-intensive services makes it theoretically possible to reach users in a very large number of countries and therefore increase the scope for geographical network spread of internationalization. Moreover, the new mode of delivery may pave the way for full FDI and thus the dynamic sequence export, and FDI could be replaced by e-transmission and FDI. For example, a positive experience with the processing of downloadable products from the USA or the UK to Ireland may lead the investing TNCs to expand their activities in Ireland and thus may lead to further FDI by the same firms or, indeed, by others.

## INDICES OF INTERNATIONALIZATION

The rest of the chapter will present an empirical study that attempts a comparative assessment of the degree of internationalization of the world's largest TNCs operating in services and/or manufacturing. Two indices are developed and estimated corresponding to the two approaches to the cross-countries dimension discussed in the previous section. The assessment refers to one mode only: the direct production abroad mode for which we have a reasonable amount of information on the two approaches.

The first index is designed to assess the *degree of foreign projection* of the company and is constructed as the percentage of direct linkages abroad in relation to the total number of direct linkages (domestic and foreign). The linkages refer to total affiliates which include subsidiaries (with an ownership stake of at least 50 per cent) and associates (with a stake of between 10 and 50 per cent).

$$Ii = FA/TA,$$

where: $Ii$ = Internationalization index, $FA$ = foreign affiliates, = $TA$ = total affiliates. This index assesses the propensity of the firm to operate away from the home country. For any random direct linkage (be it affiliate or subsidiary) of a firm, the index assesses the probability that it is located abroad.

The second index used, the Network Spread index (NSi), is designed to take account of whether the firm operates abroad in a few or many countries and thus to assess the *degree of spread* of direct activities among the various countries of the world. The index is developed in Ietto-Gillies (1998)[14] and is arrived at as follows. Let $n$ = the number of foreign countries in which the TNC has direct linkages; $n^*$ = the number of foreign countries in which, potentially, the firm could have located direct linkages; and $NSi = n/n^*$ = Network Spread index.

Theoretically, $n^*$ could include all the countries of the world; in practice, we have taken it to be the number of countries, worldwide, which have been in receipt of foreign direct investment. This is, in fact, taken as a willingness on the part of the host country to accept inward FDI and therefore as a real possibility for the firms to invest there. We have, therefore, taken $n*$ to be the number of countries in which there is inward stock of FDI minus one, in order to exclude the home country of the TNC. From the data in UNCTAD (1997, Annex, Table B.3 ), $n*$ is equal to 178 for 1997. The actual value of $n*$ is not very relevant because the analysis which we shall be making is based on comparison of the index between countries and sectors and the actual scale of the index is not relevant. We shall also give the value of '$n$' that is the actual number of foreign countries in which the firms have direct linkages.

In our empirical work, both indices are expressed as percentage. Thus the Network Spread index measures the percentage of foreign countries in which the TNC has direct linkages in relation to the total number of foreign countries in which, potentially, it could have located affiliates. Given any randomly selected country (from those that are in receipt of world FDI) the index assesses the probability that the TNC under consideration may have located direct activities in it. The Network Spread index focuses on the spread of activities into many foreign countries and not on the 'foreignness' only, as in the internationalization index.

Two data sets were used for the study: the list of the world's 1000 largest firms by market capitalization, published in *Business Week* (*BW*), and the information on the firms' ownership trees from Dun and Bradstreet's *Who Owns Whom* (1997) (WoW). The *BW* list also provides information on the home country of the firm and the industry classification within which it operates. The 664 firms selected are those with the following characteristics: they are the ultimate firms and have subsidiaries in at least one foreign country.[15]

## THE RESULTS FOR THE WORLD'S LARGEST TNCs

The empirical work[16] presented in Tables 2.1, 2.2 and 2.3 groups the firms according to whether they belong to the manufacturing or services sectors following the firms' sectoral allocation in *Business Week*. Table 2.1 gives the number and percentages of firms by home country and major sector. It shows that by far the biggest number of the world's largest TNCs are in manufacturing: 410, or 61 per cent.

As regards the countries' participation in the 664 TNCs, the USA has by far the largest number (259, or 39 per cent). Japan follows with 122 (18 per cent) firms and the UK with 88 (13 per cent). The latter country's high

*Table 2.1   World's largest 664 TNCs: breakdown by home country and
           sector, 1997*

| Home country | Total TNCs | | Manufacturing | | Services | |
| --- | --- | --- | --- | --- | --- | --- |
| | (no.) | (%) | (no.) | (%) | (no.) | (%) |
| USA | 259 | 39 | 176 | 68 | 83 | 32 |
| Japan | 122 | 18 | 82 | 67 | 40 | 33 |
| UK | 88 | 13 | 39 | 44 | 49 | 56 |
| Germany | 38 | 6 | 26 | 68 | 12 | 32 |
| France | 29 | 4 | 17 | 59 | 12 | 41 |
| Canada | 22 | 3 | 15 | 68 | 7 | 32 |
| Sweden | 19 | 3 | 12 | 63 | 7 | 37 |
| Netherlands | 13 | 2 | 6 | 46 | 7 | 54 |
| Australia | 13 | 2 | 7 | 54 | 6 | 46 |
| Switzerland | 12 | 2 | 6 | 50 | 6 | 50 |
| Hong Kong | 10 | 1 | 4 | 40 | 6 | 60 |
| Spain | 9 | 1 | 3 | 33 | 6 | 67 |
| Italy | 7 | 1 | 3 | 43 | 4 | 57 |
| Denmark | 7 | 1 | 4 | 57 | 3 | 43 |
| Singapore | 6 | 1 | 1 | 17 | 5 | 83 |
| Belgium | 6 | 1 | 3 | 50 | 3 | 50 |
| Ireland | 3 | 0 | 1 | 33 | 2 | 67 |
| Norway | 2 | 0 | 2 | 100 | 0 | 0 |
| New Zealand | 2 | 0 | 1 | 50 | 0 | 50 |
| Finland | 2 | 0 | 2 | 100 | 0 | 0 |
| Total | 669 | 98 | 410 | 61 | 259 | 39 |

participation is quite remarkable given the size of its domestic economy.
These three countries are represented in most industries both in services
and in manufacturing (Table 2.2, first three rows).

Manufacturing TNCs figure in all the listed countries, while services
figure in all the countries but two: Norway and Finland. However, within
services, two industries (banking and telecommunications) are in the portfolio of most countries. This is the case for only one sector in manufacturing (energy sources), if we ignore multi-industry which is by its nature a
hybrid and therefore likely to include both manufacturing and services
industries.

Table 2.3 gives the values and ranking for the two indices discussed in the
previous section and for the various industries within manufacturing and
services, respectively. The manufacturing sector exhibits the highest values
for the two indices: 58·4 for the internationalization index ($Ii$) and 14·3 for

*Table 2.2  World's largest 664 TNCs by industry and home country, 1997*

| | 1 Energy sources | 2 Utilities: electrical & gas | 3 Building materials and components | 4 Chemicals | 5 Forest products and paper | 6 Metals: non ferrous | 7 Metals: steel | 8 Misc. materials and commodities | 9 Aerospace and military technology | 10 Construction and housing | 11 Data processing and reproduction | 12 Electrical and electronics | 13 Electronic components and instruments | 14 Energy equipment | 15 Industrial components | 16 Machinery and engineering | 17 Appliances and household durables | 18 Automobiles | 19 Beverages and tobacco | 20 Food and household products | 21 Health and personal care | 22 Recreation, other consumer goods | 23 Textiles and apparel | 24 Multi industry | 25 Gold mines | Manufacturing: subtotal | 26 Broadcasting and publishing | 27 Business and public services | 28 Leisure and tourism | 29 Merchandising | 30 Telecommunications | 31 Transportation: airlines | 32 Transportation: road and rail | 33 Transportation: shipping | 34 Wholesale & international trade | 35 Banking | 36 Financial services | 37 Insurance | 38 Real estate | Services: subtotal | Total |
|---|---|---|---|---|---|---|---|---|---|---|---|---|---|---|---|---|---|---|---|---|---|---|---|---|---|---|---|---|---|---|---|---|---|---|---|---|---|---|---|---|---|
| **Home country** | | | | | | | | | | | | | | | | | | | | | | | | | | | | | | | | | | | | | | | | | |
| USA | 16 | 11 | 1 | 14 | 5 | 3 | 5 | 4 | 7 | | 9 | 7 | 17 | 6 | 7 | 5 | 3 | 4 | 7 | 17 | 17 | 5 | 1 | 9 | 1 | 176 | 6 | 14 | 6 | 8 | 8 | 2 | 4 | | 5 | 18 | 6 | 6 | 10 | 83 | 259 |
| Japan | 2 | 2 | 3 | 6 | 1 | 2 | 5 | 1 | | 6 | 3 | 5 | 8 | | 7 | 8 | 5 | 5 | 2 | 3 | 5 | 3 | | 5 | | 83 | 1 | 3 | 6 | 1 | 4 | 2 | 2 | | 5 | 14 | 6 | 3 | 2 | 40 | 122 |
| UK | 5 | 4 | 2 | 2 | | 1 | 1 | 1 | 2 | | | 1 | 1 | | 1 | | | | 1 | 3 | 3 | 2 | | 1 | | 39 | 3 | 7 | 6 | 8 | 4 | 1 | | 1 | | 7 | 4 | 6 | 2 | 49 | 88 |
| Germany | 4 | | 1 | 4 | | | 2 | 1 | | | | 1 | 1 | | 3 | 3 | 2 | 3 | | 1 | 1 | 1 | | 1 | | 26 | 1 | 1 | | | 1 | 1 | | | | 6 | 2 | 2 | | 12 | 38 |
| France | 2 | 2 | | 2 | | | | | | | 2 | 2 | | 2 | | | | | | | 2 | 2 | | | | 17 | 1 | 3 | | | | | | | | 4 | 1 | | | 12 | 29 |
| Canada | 2 | 2 | | 2 | 4 | 1 | | | | | 3 | 1 | 1 | | 1 | | | | 1 | 1 | 1 | | | 1 | 2 | 15 | 1 | | | 3 | 1 | | | | | 5 | | | | 7 | 22 |
| Sweden | 1 | | | | 2 | | | | | | | 1 | | | | 2 | 1 | | | | | | | | | 12 | | | | | 1 | | | | | 4 | 1 | 1 | | 7 | 19 |
| Netherlands | | 1 | | 1 | | | | | | | | | | | | | | | 1 | 1 | | | | | | 6 | 2 | | | | | | | | | 1 | 1 | 1 | | 7 | 13 |
| Australia | 2 | 1 | 1 | | | 2 | | | | | | | | | | | | | 1 | | 1 | | | | | 7 | 1 | | | | | | | | | 5 | | | | 6 | 13 |
| Switzerland | | | | | 1 | 1 | | | | | | | | | | | | | | | 2 | 1 | 2 | | | 6 | | 1 | | | 1 | | | | | 3 | 1 | | | 6 | 12 |
| Hong Kong | 1 | | | | | | | | | | | | | | | | | | | | | | | | | 4 | | | | | 1 | 1 | | | | | | 5 | | 6 | 10 |
| Spain | 1 | 2 | | | | | | | | | | | | | | | | | | | | | | | | 3 | | | | 5 | | | | | | 5 | | | | 6 | 9 |
| Italy | 1 | | | | | | 1 | | | | | | 1 | | | | | 1 | | | | | | | | 3 | | | | | 1 | | | 1 | | 2 | 2 | | | 4 | 7 |
| Denmark | | | | | | | | | | | | 1 | 1 | | 1 | | | | 1 | 1 | 1 | | | 1 | | 4 | | | | 1 | 1 | 1 | | | | 1 | | | | 3 | 7 |
| Singapore | | | | | | | | | | | | | | | | | | | | | | | | | | 1 | | | | | | | | | | 2 | | | 1 | 5 | 6 |
| Belgium | 1 | | 1 | | | | | | | | | | | | | | | | | | | | | | | 3 | | 1 | | | | | | | | 3 | | | | 3 | 6 |
| Ireland | | 1 | | | | | | | | | | | | | | | | | | | 1 | | | | | 1 | | | | | | | | | | | | | | 2 | 3 |
| Norway | 1 | | | | | | | | | | | | | | | | | | | | | | | 1 | | 2 | | | | | | | | | | | | | | 0 | 2 |
| New Zealand | | | | 1 | | | | | | | | | | | | | | | | | | | | | | 1 | | | | | 1 | | | | | | | | | 1 | 2 |
| Finland | | | 1 | | | | | | | | | | | | | | | | | 2 | | | | | | 2 | | | | | | | | | | | | | | 0 | 2 |
| Total | 34 | 27 | 12 | 30 | 11 | 12 | 10 | 6 | 9 | 7 | 12 | 20 | 28 | 6 | 18 | 20 | 10 | 15 | 16 | 27 | 36 | 14 | 1 | 26 | 3 | 411 | 16 | 30 | 12 | 22 | 21 | 8 | 6 | 2 | 6 | 82 | 18 | 26 | 10 | 259 | 669 |

47

*Table 2.3    The World's largest 664 TNCs, by sector and industry: network of affiliates – indices and ranking, 1997*

| Sectors | Ii (%) | Ranking | NSi (%) | Ranking |
|---|---|---|---|---|
| *Manufacturing* | | | | |
| Energy sources | 42.3 | 23 | 14.6 | 12 |
| Utilities: electrical & gas | 28.5 | 25 | 5.0 | 24 |
| Building materials and components | 58.0 | 15 | 10.9 | 17 |
| Chemicals | 65.0 | 11 | 19.9 | 2 |
| Forest products and paper | 49.8 | 20 | 10.7 | 18 |
| Metals: non-ferrous | 55.0 | 19 | 10.7 | 19 |
| Metals: steel | 43.7 | 22 | 12.8 | 14 |
| Misc. materials and commodities | 66.7 | 10 | 15.5 | 11 |
| Aerospace and military technology | 31.4 | 24 | 9.9 | 21 |
| Construction and housing | 45.9 | 21 | 5.4 | 23 |
| Data processing and reproduction | 79.8 | 1 | 18.9 | 4 |
| Electrical and electronics | 69.2 | 5 | 17.7 | 5 |
| Electronic components and instruments | 69.5 | 4 | 9.6 | 22 |
| Energy equipment | 67.2 | 9 | 16.9 | 6 |
| Industrial components | 67.2 | 8 | 11.8 | 16 |
| Machinery and engineering | 55.3 | 17 | 12.0 | 15 |
| Appliances and household durables | 68.4 | 6 | 16.3 | 7 |
| Automobiles | 55.9 | 16 | 16.1 | 8 |
| Beverages and tobacco | 58.3 | 14 | 16.1 | 8 |
| Food and household products | 68.0 | 7 | 20.2 | 1 |
| Health and personal care | 71.2 | 2 | 19.0 | 3 |
| Recreation, other consumer goods | 63.9 | 12 | 13.1 | 13 |
| Textiles and apparel | 70.6 | 3 | 10.1 | 20 |
| Multi-industry | 55.1 | 18 | 15.7 | 10 |
| Gold mines | 60.9 | 13 | 3.4 | 25 |
| *Total manufacturing* | *58.4* | | *14.3* | |
| *Services* | | | | |
| Broadcasting and publishing | 39.8 | 8 | 10.6 | 6 |
| Business and public services | 55.2 | 3 | 11.0 | 5 |
| Leisure and tourism | 35.9 | 9 | 8.7 | 8 |
| Merchandising | 28.7 | 11 | 4.4 | 12 |
| Telecommunications | 29.7 | 10 | 7.9 | 10 |
| Transport: airlines | 45.7 | 7 | 15.3 | 2 |
| Transport: road and rail | 25.6 | 13 | 4.1 | 13 |
| Transport: shipping | 73.5 | 1 | 15.2 | 3 |
| Wholesale and international trade | 59.0 | 2 | 18.7 | 1 |
| Banking | 46.9 | 6 | 10.2 | 7 |
| Financial services | 53.5 | 4 | 8.6 | 9 |
| Insurance | 49.2 | 5 | 11.6 | 4 |
| Real estate | 27.8 | 12 | 4.4 | 11 |
| *Total services* | *43.9* | | *9.6* | |

the Network Spread index (*NSi*), against 43·9 and 9·6, respectively, for services.

The following patterns emerge for manufacturing. The industries with a high foreign projection and with a spread of activities in a very large number of countries are mainly those in consumer products: data processing and reproduction; electrical and electronic products; appliances and household durables; food and household products; health and personal care products. The inference from these results is that in these industries internationalization is driven by market-seeking strategies on the part of the firms.

Low levels for both indices are shown by utilities; forest products and paper; metals; aerospace and military; as well as construction and housing. These are all industries for which production is very location-specific and the location is mainly the home country, with a few foreign ones. A variety of reasons are likely to lead to these results, ranging from resource-seeking strategies (forestry and metals) to political (aerospace and military), to the competitive advantages of the home countries' own firms for the construction industry, which tends to be dominated by medium and small local firms.

Some industries have a high foreign projection (high *Ii*) but the foreign activities tend to be concentrated in a few foreign countries. They tend to be industries producing machinery and equipment (electrical components and instruments; industrial components; textiles and apparel). Gold mines activities figure as high on foreign content because the firms involved in them operate from countries other than the ones where the mines are located. On the other hand, the number of countries where the mines are located is very low, which explains the low level of *NSi* (3·4 per cent).

Thus the results are the outcome of: firms' strategies; the distribution of ownership of resources by country of origin of the TNC; the TNCs' advantages vis-à-vis local firms; political/military considerations.

What about the pattern for the corresponding results for services? Table 2.3 shows high levels for both indices in the following industries: transport (shipping and airlines); business and public services; wholesale and international trade, as well as banking and insurance. All these are consumer products or have a high level of consumer product elements. They therefore exhibit the pattern of internationalization strategies that are market-seeking.

Low levels for both indices are shown by the following service industries: transportation by road; leisure and tourism; merchandising; telecommunications; and real estate. These are industries which are very location-specific and tend to be bound to the domestic arena. The exception is leisure and tourism, for which internationalization takes different forms from the direct production one. Financial services give results of high foreign projection and low spread by nation state: an indication of strategies of direct production that are highly international but concentrate in a few foreign countries.

Overall, these results show that services are less internationalized than manufacturing and that the firms in these two main sectors of the economy tend to follow similar strategies in the context of their specific industry. Nonetheless, there are limitations to a study of this nature and in particular the following. First, the classification system takes no account of new complementarities between services and manufacturing products and industries. The high service content of many material goods cannot be reflected in the results. Neither can the fact that many firms classified as manufacturing are increasingly heavily involved in services.

Second, as the study refers to one year only, we are unable to see dynamic changes due to the new technologies and/or new patterns of internationalization. Third, the study refers to one particular type of internationalization mode – foreign direct investment – and indeed not to the amount of investment but to the presence of affiliates in a foreign country. Neither traditional modes, such as trade or licensing, nor new ones are taken account of in the indices and this is likely to bias the results. Licensing can indeed be very relevant in services while the e-transmittable mode is becoming increasingly relevant for information-intensive services. Thus the overall degree of internationalization of services may be underestimated by the figures based on FDI only. In particular, these figures may fail to capture the full impact of the new technologies.

## IMPLICATIONS AND CONCLUSIONS

Any study of services, including those related to their internationalization pattern, is confronted by our difficulty in conceptualizing services and in classifying products, firms and industries within services or manufactures. The new information and communication technologies have created new products, production processes, industries and modes of delivery for products. They have also strengthened the linkages and complementaries between manufacturing and services products and industries.

Our conceptual frameworks in relation to services are still largely 'assimilationist' (Miozzo and Miles, Chapter 1 in the present volume) and this means that we attempt to fit services within the conceptual scheme of manufacturing for both innovation and internationalization. This conceptualization sees services as laggards in innovation and productivity, as well as constrained in the delivery modes as regards internationalization.

However, the new technologies and the changes they are bringing about have made these frameworks obsolete and we must start considering services in their own right and within their specific conceptual frameworks regarding innovation (Howells, 2001), internationalization and the linkages between the two. Indeed, services can now utilize a new delivery mode that affects the mode and degree of internationalization. Moreover, services are the key adopters of the new technologies and this makes them central to the current technological revolution (Barras, 1986). With the increased complementarity between the two main sectors of the economy (Gibbs, 1988) appropriate infrastructures in both sectors may be needed for development and growth (Miozzo and Soete, 2001). Thus the debate about international comparative advantages in services versus manufacturing may become increasingly misplaced: they are both necessary.

The chapter began with a discussion of two specific demarcation criteria for services and manufacturing, the first one based on the tangibility/materiality of the product and the second one on the sectoral contribution to productivity. Issues of innovation and internationalization are closely linked to these demarcations. Specifically, it was argued that the new technologies of information and communication are affecting the divide between manufacturing and services; creating new complementarities between products, between production processes and between industries; generating new modes of internationalization; leading to new products and components; and leading to an electronic-age 'new international division of labour'. Most of these processes are at the initial stage and they are likely to become more relevant in future years as the diffusion of ICTs and their effects expand.

The discussion on the degree of internationalization started with an analysis of different theoretical frameworks for the assessment of the cross-countries dimension of internationalization. Two indices were presented and estimates were given for the 664 world's largest TNCs in manufacturing and services. By far the largest percentage of these companies (61 per cent) is classified as manufacturing; however, banking and telecommunications are in the portfolio of all the countries which are home to the world's largest 664 companies in the study.

The results show that, in terms of affiliates' presence in host countries, services tend to be less internationalized than manufactured products. The firms in both manufacturing and services sectors appear to follow similar strategies whenever they deal with consumer products.

The overall conclusions are the following. The new technologies are generating enhanced scope for internationalization including a new mode: via e-transmission of the products. They are also generating new complementarities between manufacturing and services. There may also be complementaries between delivery modes as, for example, FDI may follow an initial penetration via electronic transmission of service components. The traditional classification between the three main sectors of the economy, and in particular the demarcation between manufacturing and services, may no longer be fully adequate for understanding the following features of the economic system: its productivity levels and changes.

The new technologies are generating strong specificity for services in both the innovation and internationalization fields and indeed even more in the interface between the two. This means that we can no longer use for services the conceptual frameworks developed for manufacturing. Accordingly, more research is needed on the sources of productivity growth in the economy and on the impact of the ICTs on productivity, internationalization processes modes and degrees, and the impact on the international division of labour.

We also need to develop statistics that take account of the ICT intensity of products and processes on the industrial classification side, while on the side of international data we need statistics on the various delivery modes, including the electronic ones.

## NOTES

1. This work was developed in the context of the EC-funded TSERV AITEG project, contract HPSE-CT-1999-00043. I am grateful for the support received for the empirical work.
2. These issues are further developed in Ietto-Gillies (2001, particularly chs 1, 2 and 9).
3. Of course they all use factor services, but here I am concerned with product services only.
4. Wilkins (1998, p.22) writes: 'The obscurity of the line between industry and services was recognized when in 1995, *Fortune* abandoned its separate lists of manufacturing and services companies'.
5. Technology and mechanization have spillover effects from manufacturing to agriculture which eventually lead to high productivity and low employment in the latter sector.
6. His analysis seems to ignore the social nature of work which can overcome the diminishing returns of a single person's brain.
7. Evangelista and Savona (2002) in a study of data from the Community Innovation Survey II for Italy, find that the ICTs have a negative effect on employment.
8. Jorgenson (2001) acknowledges the productivity gains from the new technologies and points out how changes in the quality of both inputs and outputs also require better statistics to allow economists a proper analysis of their impact.
9. Quah (1997) points out that successful economies are increasingly weightless economies in which immaterial products absorb an increasing share of output and growth. His weightless products include all services rather than just ICT-intensive products. However, he finds that some successful economies show a rising emphasis on IT. He also calls for a revision of the standard industrial classification.
10. Roberts (1998) denotes the latter two as 'transhuman exports' and 'wired exports,' respectively.
11. These arguments are developed more extensively in Ietto-Gillies (2001) and, to some extent, in Ietto-Gillies (2000).

12. Castellani (2001) talks of 'learning-by investing' processes.
13. Zanfei (2000) finds that a direct entry mode facilitates future external linkages with local businesses. This points to a complementary rather than a substitution evolutionary relationship between different entry modes.
14. A brief discussion of the framework for the Network Spread index is also in UNCTAD (1998, Box II.2, pp.43-4); see also UNCTAD (2001, pp.103-4).
15. Though these firms are chosen from the *BW* world largest 1000, they may not be the largest 664 because of these inclusion criteria. The list of firms is available from the author on request. It should be noted that five of these firms (ABB, RTZ, Shell, Reed and Unilever) have headquarters in two countries and thus appear in both whenever the analysis refers to the countries of origin. This is why the total number of firms appears as 669 in Tables 2.1 and 2.2.
16. I am grateful to Peter Antonioni and Marion Frenz for their support with the empirical work.

# REFERENCES

Barras, R. (1986), 'Towards a theory of innovation in services', *Research Policy*, **15** (4), 161-73.

Bhagwati, J.N. (1984), 'Splintering and Disembodiment of Services and Developing Nations', *The World Economy*, **7** (2), 133-44.

Brynjolfsson, E. and L.M. Hitt (2000), 'Beyond Computation: Information Technology, Organizational Transformation and Business Performance', *The Journal of Economic Perspectives*, **14** (4), 23-48.

Cantwell, J. (1989), *Technological Innovation and Multinational Corporations*, Oxford: Blackwell.

Cantwell, J. (1994), 'The relationship between international trade and international production', in D. Greenaway and L.A. Winters (eds), *Surveys of International Trade*, Oxford: Blackwell, pp.303-28.

Cantwell, J. (1995), 'The globalization of technology: what remains of the product cycle model?' *Cambridge Journal of Economics*, **19** (1), 155-74.

Castellani, D. (2001), 'Technological learning and exploitation of economies of scale in Italian multinationals', paper presented at AITEG Project Workshop, 'The Impact of innovation and globalization in Europe', Madrid, Universidad Complutense, 25-6 May.

Clark, C. (1940), *The Conditions of Economic Progress*, London: Macmillan.

Clegg. J. (1993), 'Investigating the determinants of service sector foreign direct investment', in H. Cox, J. Clegg and G. Ietto-Gillies (eds), *The growth of global business*, London: Routledge, pp.85-104.

Coase, R.H. (1937), 'The nature of the firm', *Economica*, **IV**, 386-405; reprinted in G.J. Stigler and K.E. Boulding (eds), *Readings in Price Theory*, London: Allen and Unwin, pp.331-51.

Cowling, K. and R. Sugden (1987), *Transnational Monopoly Capitalism*, Brighton: Wheatsheaf.

Dalum, B., C. Freeman, R. Simonetti, N. von Tunzelman and B. Verspagen (1999), 'Europe and the information and communication technologies revolution', in J. Fagerberg, P. Guerrieri and B. Verspagen (eds), *The Economic Challenge for Europe: Adapting to Innovation-Based Growth*, Cheltenham, UK and Northampton, MA, USA: Edward Elgar, pp.106-29.

Dicken, P. (1998), *Global Shift. Transforming the World Economy,* 3rd edn, London: Paul Chapman Publishing Company.

Dun and Bradstreet (1997), *Who Owns Whom. World-Wide Corporate Structure*, CD-ROM, 3rd quarter, High Wycombe: Dun and Bradstreet Ltd.

Dunning, J.H. and R. Pearce (1985), *The World's Largest Industrial Enterprises 1962–1985*, Aldershot: Gower.

Enderwick, P. (1992), 'The Scale and Scope of Service Sector Multinationals', in P.J. Buckley and M. Casson (eds), *Multinational Enterprises in the World Economy: Essays in Honour of John Dunning,* Aldershot, UK and Brookfield, US: Edward Elgar, pp.134–52.

Evangelista, R. and M. Savona (2002), 'The Impact of Innovation on Employment in Services: Evidence from Italy', *International Review of Applied Economics*, **16** (3) 309–18.

Fisher, A.G.B. (1939), 'Production, Primary, Secondary and Tertiary', *The Economic Record*, **XV**, June, 24–38.

Fröbel, F., J. Heinricks and O. Kreye (1980), *The New International Division of Labour,* Cambridge and Paris: Cambridge University Press and Editions de la Maison des Sciences de l'Homme.

Gibbs, M. (1988), 'Continuing the International Debate on Services', *Journal of World Trade Law*, **22** (3), 199–218.

Gordon, R.J. (2000), 'Does the 'New Economy' Measure up to the Great Inventions of the Past?', *The Journal of Economic Perspectives,* **14** (4), 49–74.

Gray, H.P. (1990), 'The role of services in global structural change', in A. Webster and J.H. Dunning (eds), *Structural Change in the World Economy*, London: Routledge.

Grubel, H.G. (1987), 'All Traded Services are Embodied in Materials or People', *World Economy*, **10** (3), 319–30.

Hill. T.P. (1977), 'On Goods and Services', *Review of Income and Wealth*, **23** (4), 315–88.

Howells, J. (2001), 'The Nature of Innovation in Services', in OECD (ed.), *Innovation and Productivity in Services. Industry, Services and Trade,* Paris: OECD, pp.57–79.

Hymer, S.H. ([1960] 1976), *The International Operations of National Firms: A Study of Direct Foreign Investment,* Cambridge, MA: MIT Press (published 1976).

Ietto-Gillies, G. (1992), *International Production. Trends, Theories, Effects,* Cambridge: Polity Press.

Ietto-Gillies, G. (1998), 'Different conceptual frameworks for the assessment of the degree of internationalization: An empirical analysis of various indices for the top 100 transnational corporations', *Transnational Corporations,* **7** (1), 17–39.

Ietto-Gillies, G. (2000), 'What Role for Multinationals in the New Theories of International Trade and Location?', *International Review of Applied Economics,* **14** (4), 413–26.

Ietto-Gillies, G. (2001), *Transnational Corporations. Fragmentation Amidst Integration*, London: Routledge.

John, R., Ietto-Gillies, G., Cox, H. and N. Grimwade (1997), *Global Business Strategy*, London: International Thomson Business Press.

Jorgenson, D.W. (2001), 'Information Technology and the U.S. Economy', *The American Economic Review,* **91** (1), 1–32.

Kaldor, N. (1967), 'The Role of Increasing Returns to Industry', *Strategic Factors in Economic Development*, New York: New York State School of Industrial and Labour Relations, pp.3–23.

Kitson, M. and J. Michie (1996), 'Britain's Industrial Performance since 1960: Underinvestment and Relative Decline', *The Economic Journal*, **106** (434), 196–212.

Mallampally, P. and Z. Zimny (2000), 'Foreign direct investment in services: trends and patterns', in Y. Aharoni and L. Nachum (eds), *Globalization of Services. Some implications for theory and practice*, London: Routledge, pp.25–51.

Miles, I. (1993), 'Services in the New Industrial Economy', *Futures*, **25** (6), 653–72.

Miozzo, M. and L. Soete (2001), 'Internationalization of Services: A Technological Perspective', *Technological Forecasting and Social Change*, **67** (2), 159–85.

Nayyar, D. (1988), 'The political economy of international trade in services', *Cambridge Journal of Economics,* **12** (2), 279–98.

Oliner, S.D. and D.E. Sichel (2000), 'The Resurgence of Growth in the Late 1990s: Is Information Technology the Story?', *The Journal of Economic Perspectives,* **14** (4), 3–22.

Palmer, R. (2001), *Historical Patterns of Globalization. The Growth of Outward Linkages of Swedish Transnational Corporations, 1890–1990s*, Stockholm: Almqvist & Wiksell International.

Peoples, J. and R. Sugden (1991), 'Divide and rule by transnational corporations', in C.N. Pitelis and R. Sugden (eds), *The Nature of the Transnational Firm,* 2nd edn, London: Routledge, pp.174–92.

Preissl, B. (1995), 'New Landscapes in the Service Sector Information & Communication Technologies & Structural Change', Research Papers Series, Aston Business School Research Institute, Aston University.

Quah, D.T. (1997), 'Increasing weightless economies', *Bank of England Quarterly Bulletin,* **37** (1), February, 49–56.

Roberts, J. (1998), *Multinational Business Service Firms. The Development of Multinational Organizational Structures in the UK Business Services Sector,* Aldershot: Ashgate.

Sapir, A. and C. Winter (1994), 'Services Trade', in D. Greenaway and L.A. Winters (eds), *Survey in International Trade*, Oxford: Blackwell.

Solow, R.M. (1987), 'We'd Better Watch Out', *New York Times Book Review*, 12 July.

Stern, R.M. and B.M. Hoekman (1987), 'Negotiations on Services', *The World Economy*, **10** (1), 39–60.

Stigler, G.J. (1951), 'The division of labor is limited by the extent of the market', *The Journal of Political Economy,* **LIX** (3), 185–93.

Sullivan, D. (1994), 'Measuring the Degree of Internationalization of a Firm', *Journal of International Business Studies,* **25** (2), 325–42.

Tomlinson, M. (2001), 'A New Role for Business Services in Economic Growth', in D. Archibugi and B-Å. Lundvall (eds), *The Globalizing Learning Economy*, Oxford: Oxford University Press, pp.97–107.

United Nations Conference on Trade and Development (1997), *World Investment Report 1997. Transnational Corporations, Market Structure and Competition Policy,* Geneva: United Nations.

United Nations Conference on Trade and Development (1998), *World Investment Report 1998. Trends and Determinants,* Geneva: United Nations.

United Nations Conference on Trade and Development (2001), *World Investment Report 2001. Promoting Linkages,* Geneva: United Nations.

United Nations Conference on Trade and Development – Division on Transnational Corporation and Investment (1995), *World Investment Report 1995. Transnational Corporations and Competitiveness,* Geneva: United Nations.

Verdoon, P.J. (1949), 'Fattori che regolano lo sviluppo della produttivita' del lavoro', *L'Industria*, no. 1, 3–11.
Wilkins, M. (1998), 'Multinational corporations: an historical account', in R. Kozul-Wright and R. Rowthorn (eds), *Transnational Corporations and the Global Economy,* London: Macmillan, pp.95–133.
Williamson, O.E. (1975), *Markets and Hierarchies: Analysis and Antitrust Implications*, New York: Free Press.
Williamson, O.E. (1981), 'The economics of organisation: the transaction cost approach', *American Journal of Sociology*, **87** (November), 548–77.
Zanfei, A. (2000), 'Transnational firms and the changing organization of innovative activities', *Cambridge Journal of Economics*, **24** (5), 515–42.

PART II


Technology and trade and foreign investment
in services: a statistical appraisal

# 3. The internationalization of European services: what can data on international services transactions tell us?

## Paul Baker, with Marcela Miozzo and Ian Miles

### INTRODUCTION

This chapter examines the trends and patterns in trade and foreign direct investment (FDI) among services in the European Union (EU).[1] The continuing liberalization of services, the creation of the EU internal market and, more recently, the introduction of the single currency are adding to the pressures restructuring service sectors.

Around the world, the competitive environment for service firms is changing. As barriers to cross-border trade and foreign investment are reduced, firms can compete in wider markets, outside their home location. With growing market size, the possibilities for economies of scale and scope also grow. Combined with the internationalization of client industries, these changes are resulting in growth in services trade and investment on a world scale.

These developments in the international environment, together with the important developments in new information and communication technologies, suggest an intensification in the process of internationalization of services. But so far neither the mechanisms by which this is achieved nor the approaches adopted by individual firms have been explored in detail. This chapter examines the available statistical evidence on international services transactions. The first section describes the developments underlining the increasing internationalization of services. The second section classifies the different types of international services transactions and the available data sources. The third section explores the implications of internationalization of services for growth and competitiveness. The fourth section analyses the importance and evidence of trade in services. The fifth section analyses the importance and evidence of foreign direct investment in services. The sixth section analyses two dimensions to the international integration of services: the integration of service markets within the EU and the integration of the EU in worldwide services markets. A final section presents some conclusions from these analyses.

## FACTORS PROMOTING THE INTERNATIONALIZATION OF SERVICES

The internationalization of services is nothing new, but we are seeing an intensification of this phenomenon. A number of specific developments in the international environment underlie this. These include state retrenchment, regulatory reform, liberalization and the accompanying changes to the competition environment, and the new possibilities for coordination offered by the use of new information and communication technologies (ICTs). These developments are, of course, frequently cited in the internationalization of goods-producing sectors; they are equally relevant to the internationalization of services.

Regulatory reforms in such sectors as financial services, air transport and telecommunications have promoted the international expansion of firms within these sectors, but it has also helped provide the necessary infrastructure for the internationalization of other industries. Regional initiatives towards the liberalization and deregulation of services sectors have come from the internal market programme of the EU and from new multilateral structures like NAFTA and Mercosur. Globally, negotiations at the WTO (World Trade Organisation) regarding the General Agreement on Trade in Services (GATS) aim to further liberalize markets and promote competition in services.

Prior to the completion of the EU internal market, regulations affecting the degree of competition often made it difficult for competitors from other member states to enter service markets. Typically, these regulations imposed restrictions on the conduct of firms in ways that created significant barriers to entry for foreign firms. The Treaty of Rome stipulates the freedom to provide services and to set up establishments, but this was, in practice, hindered by problems, for example in defining the equivalence of qualifications (important for many services) and in entering certain markets organized by national governments (or by private groups sanctioned to organize and protect them). In the early 1970s, rulings of the Court of Justice in principle liberalized all services connected with agriculture, craft and trade.[2] Outstanding problems were 'solved' by the programme to complete the internal market in services by liberalization (that is, free service delivery), by mutual recognition of the quality of member states' control and by minimum sets of EU regulations.

The harmonization of EU regulations sets a minimum level of rules governing some important features of the behaviour of service providers and of the control systems of member states. Other than this minimum set of rules, there is no market regulation policy for services at the EU level, other than the general rules of EU competition policy. Consequently, at a national level, many services remain heavily regulated. This has led to a situation in which firms are able to compete freely, but may be subject to different sets of rules. Governments may compete in 'optimal' rule setting. National programmes aimed at further liberalization have taken place against this background, so that the overall degree of openness of services markets has increased and new opportunities for internationalization have been created; but the degree and speed of liberalization still vary considerably across countries and services sectors.

The new ICTs have created possibilities for restructuring both production and delivery that can be seen to be increasingly influencing service providers' strategies for internationalization. In many cases ICTs have made services more tradable by enabling new forms of storage and transmission of services and new means of producer–consumer interaction. The growth of the Internet has, for example, provided a vehicle for rapid and wide-reaching delivery of certain services. Thus some services for which achieving a critical mass in a foreign market required either establishing a network of branches or acquiring an existing local network now find that the Internet provides a realistic alternative to physical local presence (for example, in banking, retail distribution, on-line news and entertainment services).

The impact of electronic commerce can be seen as even more far-reaching for some services than for goods. It provides a means not only of purchasing but also of supplying a range of business and consumer information-related services. In other words, it is no longer necessary for such services either to be embodied in goods in order to be delivered or to require direct interaction between the supplier (or his representative) and the customer for transactions (international or otherwise) to take place. Electronic commerce may also promote the internationalization of services through replacement of traditional intermediary functions (that is, local representatives may become less important) and the scope for 24-hour trading (that is, time-related geographical boundaries are eroded).[3]

More generally, ICTs have served to accelerate international transactions (for example, cross-border transmission of information, data and financial capital) and enable the automation of global management systems (for example, financial systems, transport and logistics, personal communications). With more rapid communication and transfer of information, new modes of organization have become possible. ICTs simultaneously enable more decentralized and autonomous decision making while allowing for centralization of certain strategic services (Hatzichronoglou, 1996). For business services sectors, ICTs have had a dual effect. First, they increase the demand for those services connected to the implementation and management of the technologies themselves. Secondly, as many business services are information-intensive, they lend themselves to the application of technologies that enable them to collect, analyse and disseminate information rapidly.

ICTs may also support the increased 'commoditization' or 'standardization' of certain types of services, thus making them more susceptible to economies of scale. There is relatively little analysis of the extent to which service activities are standardized: many services are generally assumed to require a high degree of specialization or customization. The preponderance of small firms in many service sectors points to a high degree of heterogeneity in service outputs. There are exceptions, however, in banking and finance, transport, communications and some retail segments. Standardized products are common where (average) firm size tends to be higher. For network services (for example, banks, insurance and telecommunications), as opposed to scale-intensive services (transport, wholesale trade and distribution), ICTs have facilitated the complexity, precision and quality of services and may actually increase the possibility for customization (Soete and Miozzo, 1989).[4]

Some degree of standardization is implicit in the internationalization of services – the products themselves, or at least elements of their production, need to be replicated for sale in foreign markets (Rubalcaba-Bermejo, 1999). Where ICTs increase the possibility for standardization, they should also increase the tradability of services. At the same time, standardization can make for greater economies of scale, and thus, potentially, for more price-based competition. The search for larger markets and economies of scale, and the response to overseas firms entering previously sheltered markets, provide incentives for firms to internationalize their activities.

The changes to the international environment brought about by regulatory changes and new ICTs provide a basis for explaining the acceleration of the process of internationalization of services. It is not difficult to see why the dismantling of barriers to competition in service sectors, such as telecommunications and air transport, has resulted in firms seeking to reinforce their international position through mergers and acquisitions, or via alliances and cooperation agreements. For firms that seek to provide services to clients who themselves have, or seek, a wide geographical reach, there is risk in being absent from any significant segment of the world market. The opening up of markets gives potential competitors more scope to exploit the absence of a firm from a particular market segment (geographical or otherwise) and to use such an entry point to challenge an incumbent supplier. Ultimately, such competitors may threaten the firm's own 'home' market. New ICTs that complement the ability of competitors to enter new or previously closed market segments serve to reinforce this potential competition. We confront a rapidly evolving international environment, to which responses vary. A slower and more progressive approach to internationalization can now be seen to be too 'risky' a strategy. At the same time, firms within markets that are subject to regulatory reform or, more generally, are becoming more competitive may judge securing a strong defensible home base to be a prerequisite for future international expansion.

## TYPES OF INTERNATIONAL SERVICES TRANSACTIONS AND DATA SOURCES

One of the key characteristics of many services is that they are intangible – in the familiar saying, 'you can buy them and sell them, but you can't drop them on your foot' – and they are, thus, not tradable in a conventional sense. The service cannot (usually) be put in the freight compartments of boats, planes or lorries, stored in warehouses, stacked on counters. We are used to thinking of services as intangible things that people do, whereas goods are tangible things that people make, but it is less useful to demarcate sharply two clearly distinct groups, services (intangible) and goods (tangible), than to envisage a continuum along which outputs contain varying proportions of services and goods or tangible and intangible elements (see Goedegbuurte, 1996). Thus Dunning (1989) comments that 'the output of economic activity may range from that of pure goods to pure services. However, most (indeed an increasing proportion of) goods embody some non-factor intermediate services, and most services embody some intermediate goods. And even pure services require people to supply them'.

Consequently, trade in some services may be achieved through trade of the goods in which they may be embodied. For others, trade may require accompanying goods to be traded (or at least be available) in order for a transaction to take place. Other services, which are (or come closer to being) 'pure services', may require little or no accompanying trade in goods (although, if they are reliant on ICT for international delivery, these are necessarily goods required as infrastructure and peripherals).

From the perspective of international transactions in services, and especially the statistics dealing with such transactions, the service component embodied in many goods cannot be separately identified from the trade in goods themselves. Thus service outputs that are readily embodied in goods are excluded from the definition of services for the purposes of collecting statistics on trade in services, for example. Computer software provides one obvious example where this is the case. Essentially, computer software provides a service to the user rather than a tangible good, yet, where this service is supplied (embodied) in a physical good such as a diskette or CD-ROM, it is counted as a trade in goods for the purposes of trade statistics. It can be seen from this example that there is a built-in restriction, based on the extent to which services are embodied in goods, limiting the scope of the definition of services when it comes to the main sources of statistics on international service transactions.

The technical possibility of embodying services in goods is not the only determinant of the 'tradability' of services. A closely related feature of services, frequently identified as affecting their 'tradability', is the need for interaction (proximity) between producer and consumer for a service transaction to take place. In the strongest case, this interaction requires that both the supplier and the consumer be present for a transaction to take place – coterminality. Cross-border supply alone is, generally speaking, only an option where the required level of supplier–customer interaction is low (even though this does not necessarily mean that the frequency of interaction is low). Nonetheless, cross-border trade is important for some service sectors. This has led to various attempts to characterize international service transactions (for example, Sampson and Snape, 1985; Vandermerwe and Chadwick, 1989; Sapir and Winter, 1994). A basic distinction is drawn between services that may be traded in a conventional sense (that is, cross-border supply) and those that require factor movements (that is, movement of the supplier, either temporarily or permanently, to the location of the client).

The importance of non-conventional modes of trade for services is reflected in the General Agreement on Trade in Services. Recognizing the greater diversity of modes of service 'trade' compared to goods, GATS identified four modes of producer–consumer international interactions. Likewise, Karsenty (1999) gives an indication of the relative importance of the different modes of supply (as outlined in Table 3.1). Although the estimates should be treated with caution, they provide a rough indication of the value of services trade. His estimates indicate that trade in services amounted to $2170 billion in 1997, or roughly 30 per cent of world trade. On the basis of these estimates, modes one and two, which represent the part of total services trade covered by (conventional) trade statistics, are found to account for the largest part of trade in services. Nevertheless, some two-fifths of trade in services is via other modes.

*Table 3.1   Trade in service, by mode of supply, 1997*

| Mode of supply (with brief definition) | Statistical proxy used to obtain estimate | Value ($US billion) | Share of total services trade (%) |
|---|---|---|---|
| Mode 1: Cross-border supply A service flows from the territory of one country into the territory of another country (e.g. banking or architectural services transmitted via telecommunications or mail) | Business services shown in balance of payment statistics (excluding tourism and travel) | 890 | 41.0 |
| Mode 2: Consumption abroad A service consumer or his/her property moves into another country's territory to obtain a service (e.g. tourism, ship repair or aircraft maintenance) | Tourism and travel | 430 | 19.8 |
| Mode 3: Commercial presence A service supplier of one country establishes a territorial presence, including through ownership or lease of premises, in another country's territory to provide a service (e.g. insurance firms or hotel chains) | Foreign affiliates production (estimates of gross output) | 820 | 37.8 |
| Mode 4: Movement of personnel Persons of one country enter the territory of another country to supply a service (e.g. accountants, doctors or teachers) | Compensation of foreign employees (shown in balance of payment statistics) | 30 | 0.1 |
| All modes | | 2170 | 98.7 |

*Source:*   Karsenty (1999).

Conventional trade in services (that is, modes one and two) is important in many services: even allowing for the importance of travel and transport,[5] there is significant trade in other services sectors. But conventional trade significantly underestimates the importance of overall trade in services. Moreover, it should be noted that Karsenty adopts only a limited definition of commercial presence (that is, foreign affiliates) which may also be obtained through third party methods, such as franchising, licensing management agreements and minority joint ventures, that are important in many service sectors (certain retail segments, fast food chains, hotels and so on). There is little evidence as to the value of production that such methods of obtaining commercial presence in foreign markets generate.

One question that arises concerns the extent to which recorded (conventional) trade in services is in fact conditional on (the pre-establishment of) commercial presence. Or, similarly, to what extent does cross-border supply of services reflect intra-firm transactions, the result of franchising and other third party mechanisms, or direct supply by the service provider to the final client? In this context, it is worth noting that for many widely traded services (for example, ICT services, banking, insurance and so on) intra-firm trade is substantial. Although a number of empirical studies have examined the general relationship between exports and FDI (Fontagné, 1999), there has been little specific attention to the situation with respect to services. Given the nature of service transactions, it seems reasonable to suppose that the greater complementarity, or at least conditionality, of trade and FDI will be handled differently in the case of different services, and that assumptions derived from the analysis of manufacturing trade and investment may not generalize well to many services.

One way in which services differ is in the necessary levels of FDI required to establish a viable foreign commercial presence. In sectors such as finance these levels may be considerable, while for less capital-intensive services, the assets of which are mainly incorporated in their personnel, the cost of setting up a commercial presence (in terms of buildings and equipment) may be low. In such circumstances, the level of 'exports' necessary to offset the costs of establishing a presence in a foreign market may also be low. For this reason the complementarity and/or substitutability between (cross-border) trade and FDI may be even less well defined than for manufacturing. But it is worth noting that many of those industries that are generally recognized as being investment-intensive (for example, banking and insurance, telecommunications) also account for a high proportion of recorded cross-border trade.[6]

Relationships between FDI and trade are not limited to service-to-service interactions. There is, for example, a clear relationship between FDI in wholesaling services and trade in goods. A foreign wholesaling affiliate may be established, with the primary function of distributing products of the parent firm, in order to gain access to a foreign market – enabling the parent to achieve economies of scale not only in production but also in distribution, advertising and so on. (In trade statistics, wholesale trade services incidental to the wholesaling of merchandise are indistinguishable from merchandise (goods) trade data, which has the effect of understating significantly the importance of trade in wholesale services). In Europe, the internal market, with its reduction in intra-EU custom barriers, has greatly facilitated such scale economies. The removal of national barriers within the EU has indeed prompted consolidation within the wholesaling sector in the EU, with both intra- and extra-EU consolidation being reflected in the high levels of recent mergers and acquisitions activities in the wholesaling services sector, which is characterized by a high presence of foreign affiliates.

Analysts have been particularly interested in the behaviour of American firms, and US trade data are relatively well-developed. WTO studies indicate that, for the United States, and for trade as a whole, 'investments by foreign parent firms in their US affiliates provide a vital bilateral linkage that largely determines the size and growth of their intra-firm trade transactions' (WTO, 1998a). With respect to EU–US trade in services, trade between related parties[7] is found to be the dominant factor in total EU–US trade in services.[8] The US surplus in EU–US-related parties trade in services is mostly accounted for by royalties and licence fees, suggesting a US comparative advantage based on intellectual property, such as technology, produced by the US parents and used by their EU affiliates (ibid.).

Table 3.2 shows a clearer picture of the different modes of supply and statistical sources available. Appendix 1 describes the basic sources of statistical data, about which several reservations must be noted. These data can only provide an incomplete picture of international transactions and, hence, the internationalization of services. They provide partial information on modes one, two and three of trade in services. They do not help to identify third-party means of obtaining commercial presence, nor do they indicate the importance of mode four (movement of personnel).[9] It should also be noted that asymmetries in data on trade and FDI for services are a considerable problem, owing to methodological differences across countries and in the timing of recording of transactions.

## INTERNATIONALIZATION, GROWTH AND COMPETITIVENESS

Internationalization is not only a phenomenon that affects individual firms or sectors but one that has wider-reaching implications for regions and countries. The degree of involvement in the process of international integration of markets can be viewed as important, whatever the modalities (for example, production, export, cooperation) by which it is achieved. Successful internationalization, engendering economic growth, depends in large part on the collective ability of (incumbent or domestic) firms to

*Table 3.2    Modes of supply and statistical data sources*

| GATS mode | Conventional trade (BOP) | Other 'trade' | Investment |
|---|---|---|---|
| Cross-border supply | Trade in services (other than tourism and travel) | | |
| Consumption abroad | Trade in services (mainly tourism and travel) | | |
| Commercial presence | | | |
| Foreign affiliate (direct investment) | Intra-firm trade in services *Included in trade in services* | Foreign affiliates trade in services (FATS) | Foreign direct investment (FDI) Mergers and acquisitions (M&A) |
| Third party representation (e.g. franchise, license, joint-venture) | Inter-firm trade in services *Included in trade in services* | Royalties and licence fees (BOP) *Services not separately identified* | |
| Presence of natural persons | Trade in services (other than tourism, travel and transport) *Not separately identified from cross-border supply* | Compensation of foreign employees (BOP) | |

*Source:*    WTO 1995, 1998b; UNCTAD, 1997.

compete internationally. But it also depends on the region's ability to attract internationally successful firms or to forge partnerships with them. Thus the potential benefits (and risks) for countries and regions may result from outward internationalization (that is, by home based firms) and/or from inward internationalization (that is by foreign-based firms).

A considerable economic literature exists on the benefits of trade (and foreign direct investment) much of it extolling these benefits in an uncritical manner. Services are rarely treated explicitly (see, for example, Hindley and Smith, 1984; Deardorff, 1985; Sampson and Snape, 1985; François, 1990) and are often assumed to be no different in principle from goods. The degree to which theories developed in the context of trade in goods are applicable to services is debatable. The application of the concept of comparative advantage may be relatively straightforward with respect to services that can be readily standardized and traded as identifiably separate units. For many other services, greater importance is likely to attach to commercial presence and, hence, to FDI. For analytic purposes, then, the task may therefore be to determine the extent to which the characteristics of a given service are sufficiently close to those of goods for the same theoretical considerations to apply, or so different that arguments about the implications of FDI for both host and source economies are more appropriate.

FDI, whether through mergers and acquisitions (M&A) or through greenfield investments, has important implications for both home and, particularly, host economies, as it affects the control and characteristics of economic activity in ways different from cross-border trade. A distinction is made between production and financial implications of FDI (see Stevens and Lipsey, 1992). The main financial implications of FDI relate to possible substitution between foreign and domestic investment, either because investing abroad may reduce investment in a firm's home location or because foreign investment crowds out domestic investment in the host location. Thus, in either location, there is a risk that internationalization may reduce domestic investment and potentially have (longer-run) negative impacts on competitiveness. Alternatively, for the host country, foreign investment may encourage matching or complementary investments by local suppliers and partners, or stimulate investment by domestic or other foreign incumbent firms in response to increased competition.

The main direct production-related implication of FDI, as with conventional trade, is through the effect of intensified competition on domestic firms. New entrants may contest the (monopolistic) position of existing firms. This may force them to improve their competitiveness through, for example, lower prices, higher quality and innovation, and improved capital and labour productivity. Client industries and final consumers may benefit from the increased diversification of service products (quality, range and price) brought by a foreign entrant and through its impact on incumbent suppliers.

Inward investment may also result in beneficial inter- or intra-industry spillovers (Peneder, 2000). Horizontal spillovers may arise, for example, where foreign firm affiliates train the domestic workforce or improve standards of management. Vertical spillovers may involve foreign firms having impacts on downstream (for example, suppliers) and upstream (for example, distribution) segments of the production and distribution chain.

The nature of many service activities suggest that, where knowledge and technology gaps occur, they may be overcome, among other means, through domestic learning from spillovers from the owners of the technology and knowledge concerned. Many of the characteristics that are common to a range of service activities relate essentially to intangible and often non-codifiable assets (for example, the need for personal contact between producer and client, the importance of quality and reputation, human capital and the ability to retrieve and analyse information). Despite the efforts of firms to protect their knowledge-based advantages ('intellectual assets'), the fact that knowledge is only partially excludable suggests that spillovers may eventually occur.

Technology and knowledge-related spillovers associated with foreign direct investment (or other forms of collaboration and third party relationships) may have the potential to stimulate (endogenous) growth in the host economy. Spillovers may, therefore, provide a key motivation for governments to attract foreign direct investment. Moreover, where economies of scale and scope in the production of technology and knowledge are important, the potential for technology gaps to occur may be greatest for small economies. Accordingly, spillovers may be relatively more important for small countries that are able to attract foreign investment; Ireland and Singapore are cases in point.

## ANALYSING DATA ON TRADE IN SERVICES

Looking at (conventional) trade data, total trade in services as measured by the sum of exports and imports is larger for the EU than for either the USA or Japan (see Table 3.3). It is also more important for the EU when measured as a share of GDP; in 1998, total trade in services for the EU was equivalent to approximately 5·5 per cent of GDP, compared to 4 per cent for the USA and Japan. Moreover, as a percentage of GDP, trade in services has been growing more quickly for the EU than for the other Triad members. The USA, however, maintains a greater surplus in trade in services than the EU, while Japan has a large deficit. Prior to 1998, when both the USA and the EU saw a decline in their surplus in trade in services, their net trade position had been increasing. Japan's trade deficit had been increasing, although it improved somewhat in 1997 and 1998 when trade slowed as a result, at least in part, of the Asian crisis.

The EU has the highest share of trade in services to total trade (23·5 per cent): its share of (external) trade in services to total trade is of roughly the same size for exports and imports, 23·9 per cent and 23·0 per cent, respectively, in 1998. This is not true for the USA and Japan (see Table 3.4): the USA exports relatively more services (23·2 per cent of total exports of

*Table 3.3    Breakdown of services trade for Triad members in 1998 (billions of ECU)*

|  | European Union | | | United States | | | Japan | | |
|---|---|---|---|---|---|---|---|---|---|
|  | Exports | Imports | Net | Exports | Imports | Net | Exports | Imports | Net |
| Goods | 701.7 | 674.8 | 26.9 | 599.6 | 818.1 | −218.5 | 333.8 | 224.6 | 109.2 |
| Services[a] | 220.3 | 202.0 | 18.3 | 181.3 | 138.2 | 43.1 | 48.6 | 90.8 | −42.2 |
| Transport | 60.7 | 56.8 | 3.9 | 40.6 | 44.8 | −4.2 | 19.0 | 25.4 | −6.4 |
| Travel | 60.9 | 62.3 | −1.3 | 74.7 | 51.7 | 23.0 | 3.4 | 25.7 | −22.4 |
| Construction | 12.3 | 7.1 | 5.2 | 3.6 | 0.6 | 3.0 | 6.9 | 4.9 | 2.0 |
| Other services | 86.3 | 75.8 | 10.5 | 62.3 | 41.0 | 21.3 | 19.3 | 34.8 | −15.5 |

*Note:*    [a] Excluding royalties and licence fees, and government services (not included elsewhere).

*Source:*    NewCronos; NEI calculations.

*Table 3.4    Trade in services[a] as a share of total trade by Triad members in 1998 (per cent)*

|  | Exports | Imports | Total Trade |
|---|---|---|---|
| Extra-EU | 23.9 | 23.0 | 23.5 |
| United States | 23.2 | 14.5 | 18.4 |
| Japan | 12.7 | 28.8 | 20.0 |

*Note:*    [a] Excluding royalties and licence fees, and government services (not included elsewhere).

*Source:*    NewCronos; NEI calculations.

goods and service) than it imports (14·5 per cent), and Japan displays the opposite trend, with services having a 28·8 per cent share of imports in 1998 and only 12·7 per cent of exports.

Both the EU and the USA witnessed faster growth in services trade during the period 1995–8, compared to 1992–5. By contrast, the annual average growth rate decreased for both exports and imports in Japan (but this is unlikely to be a long-term trend: recall the economic and financial crisis in Asia). The growth rates for the USA were the highest for Triad members in both periods (see Table 3.5).

We suggested above that there may be strong motivations for firms in smaller countries to trade more intensively, and seek proportionally higher

*Table 3.5    Annual average growth rates of trade for the Triad (per cent)*

| | Exports | | | Imports | | |
|---|---|---|---|---|---|---|
| | 92–5 | 95–8 | 92–8 | 92–5 | 95–8 | 92–8 |
| EU15 (extra) | | | | | | |
| Goods | 10.3 | 8.9 | 9.6 | 6.3 | 8.3 | 7.3 |
| Services[a] | 4.7 | 10.2 | 7.4 | 4.3 | 10.6 | 7.4 |
| USA | | | | | | |
| Goods | 9.1 | 10.8 | 10.0 | 11.5 | 12.6 | 12.1 |
| Services[a] | 5.2 | 11.7 | 8.4 | 7.4 | 13.6 | 10.4 |
| Japan | | | | | | |
| Goods | 8.7 | 0.7 | 4.6 | 14.1 | −0.3 | 6.6 |
| Services[a] | 7.6 | 3.0 | 5.3 | 7.3 | 1.9 | 4.6 |

*Note:*   [a] Excluding royalties and licence fees, and government services (not included elsewhere).

*Source:*   NewCronos; NEI calculations.

levels of FDI. The governments of these countries are likely to create incentives, which may be either financial or through the creation of positive externalities, to encourage foreign direct investment flows. Analysis of the relative importance of trade for EU member states, whether in goods or in services, supports the hypothesis that smaller countries will trade more intensively. Specifically for trade in services, the smaller member states reveal the highest trade (exports plus imports) to GDP ratios (see Table 3.6). Five member states, Ireland, the Netherlands, Belgium and Luxembourg, and Austria display particularly high trade to GDP ratios.

Intriguingly, it would also appear that there is a relationship between size of the domestic economy and the ratio of intra-EU trade to total trade. The share of intra-EU trade is noticeably lower for the four largest European economies, indicating a greater orientation outside Europe. Spain and Portugal, by contrast, reveal the greatest orientation of services trade to the EU.

On the basis of the two indicators (trade related to GDP, and intra-EU as a share of total trade), three groupings of countries appear to arise. These are as follows:

1.  Major economies, characterized by low trade to GDP ratios and a relatively balanced orientation between extra- and intra-EU trade: Germany, France, Italy and the UK.
2.  Small/central economies, characterized by high trade to GDP ratios

*Table 3.6   The importance of trade in services for individual member states (per cent)*

|  | Trade in services/GDP | | | Service trade/ total trade | Intra-EU services trade/total services trade |
|---|---|---|---|---|---|
|  | 1992[a] | 1995 | 1998 | 1998 | 1998 |
| EU15 | 11.3 | 11.4 | 12.6 | 20.6 | 54.4 |
| Germany | 7.4 | 7.8 | 9.1 | 16.3 | 54.7 |
| France | 11.7 | 9.2 | 10.0 | 19.5 | 49.6 |
| Italy | 8.9 | 10.5 | 10.7 | 22.0 | 52.2 |
| Finland | 10.2 | 12.9 | 10.8 | 15.7 | 60.5 |
| UK | 9.6 | 11.4 | 11.5 | 21.8 | 41.2 |
| Spain | 8.7 | 10.2 | 12.7 | 23.7 | 68.0 |
| Portugal | 10.1 | 13.4 | 13.6 | 18.8 | 72.9 |
| Greece | 15.6 | 13.8 | 15.1 | 41.1 | 59.6 |
| Sweden | 12.9 | 12.8 | 15.8 | 19.8 | n.a. |
| Ireland | 18.1 | 20.0 | 23.7 | 15.9 | n.a. |
| Netherlands | 21.4 | 20.8 | 23.7 | 22.3 | 60.1 |
| Belg.–Lux. | 24.1 | 21.9 | 25.3 | 18.5 | 66.8 |
| Austria | 23.5 | 24.9 | 28.8 | 32.0 | 63.9 |

*Note:*   [a] Excluding trade in royalties and licence fees, and government services; for Greece data for 1993.

*Source:*   NewCronos, Eurostat; NEI calculations.

and a moderate orientation towards intra-EU trade: the Netherlands, Belgium–Luxembourg and Austria.

3.   Small/periphery economies, characterized by medium trade to GDP ratios and a high orientation towards intra-EU trade: Spain, Portugal, Finland and Greece.

(Unfortunately, data in the intra/extra-EU split are unavailable for Ireland and Sweden.)

## FDI IN SERVICES

FDI is more difficult to analyse, owing to limited harmonized data on, for example, the relationship between FDI (particularly FDI stocks) in services

*Table 3.7    The importance of FDI flows in services for individual member states (per cent)*

| | FDI flows in services/GDP | | | Service FDI/ total FDI | Intra-EU services FDI/total services FDI |
|---|---|---|---|---|---|
| | 1992 | 1995 | 1998 | 1998 | 1998 |
| EU15 | 1.2 | 1.7 | 3.7 | 55.5 | 56.2 |
| Italy[a] | 0.5 | 0.5 | 0.8 | 62.5 | 49.9 |
| Austria | n.a. | n.a. | 2.3 | 73.0 | — |
| Germany[b] | 0.7 | 1.4 | 2.4 | 49.9 | 60.0 |
| Portugal | 1.8 | 0.4 | 2.5 | 71.0 | 49.9 |
| France | 1.6 | 0.9 | 3.2 | 71.4 | 59.2 |
| Spain | 0.7 | 1.0 | 3.3 | 63.7 | 43.0 |
| UK | 1.3 | 1.9 | 3.6 | 35.2 | 14.0 |
| Denmark[c] | 0.9 | 3.1 | 3.8 | 63.9 | 40.5 |
| Sweden | 1.2 | 0.7 | 8.8 | 64.1 | 89.1 |
| Netherlands | 3.4 | 4.8 | 10.0 | 54.7 | 58.0 |
| Finland[b] | 0.1 | 0.1 | 14.9 | 60.1 | 97.6 |

*Notes:*
[a]  Excluding real estate and business activities for 1998 and 1995.
[b]  Excluding hotels and restaurants.
[c]  Excluding hotels and restaurants for 1998; excluding real estate and business activities, and other services for 1992.

*Source:*    NewCronos, Eurostat; NEI calculations.

and the size of the domestic economy, in the way that we did for trade. Nonetheless, it would appear that geographical and institutional factors may be influential in this case. FDI flows tend to be proportionally more important for Nordic countries, the UK and, especially, the Netherlands. But they are relatively low for Latin countries, Germany and Austria (see Table 3.7). The UK and the Netherlands also have relatively high levels of FDI stocks (assets plus liabilities) relative to GDP (see Table 3.8). However, the relatively low shares of services in total FDI indicates that it is not services that are specifically important for these countries – rather that they have high aggregate levels of FDI of all sorts relative to GDP. Clearly this is a topic that deserves more research.

*Table 3.8*   *The importance of FDI stocks$^a$ in services for individual member states (per cent)*

| | FDI position in services/GDP | | Service FDI/ total FDI | Intra-EU services FDI/total services FDI |
|---|---|---|---|---|
| | 1995 | 1997 | 1997 | 1997 |
| EU15 | 15.5 | 18.6 | 54.9 | 56.9 |
| Finland$^b$ | 3.7 | 3.9 | 15.8 | 74.8 |
| Italy$^c$ | 8.5 | 8.6 | 47.1 | — |
| Austria | n.a. | 10.6 | 69.1 | 60.5 |
| France | 11.8 | 13.0 | 54.1 | 59.9 |
| Germany | 12.9 | 16.5 | 72.6 | 57.4 |
| Portugal | 11.6 | 16.7 | 68.7 | 69.0 |
| UK | 18.3 | 22.8 | 46.3 | 41.6 |
| Netherlands | 37.7 | 46.1 | 50.2 | 57.1 |

*Notes:*
$^a$ Sum of FDI holdings of firms from reporting country (assets) and FDI holdings of foreign firms in the reporting country (liabilities).
$^b$ Excluding hotels and restaurants.
$^c$ Excluding hotels and restaurants, and real estate and other business activities; also excluding other business services in 1997.

*Source:*   NewCronos, Eurostat; NEI calculations.

## AN INTEGRATION OF SERVICES IN THE EU?

The international integration of services is relevant for the EU both in terms of the integration of service markets within the EU and in terms of the integration of the EU in global services markets. The growing importance of both intra- and extra-EU trade and, in particular, of investment flows in services relative to GDP (see Figures 3.1 and 3.2) provides evidence that both dimensions of integration are at play. In both the intra- and extra-EU case, FDI is growing more quickly than cross-border trade in services, suggesting the continuing – and possibly increasing – importance of commercial presence for international service delivery. One possibility is that there is a gradual replacement of trade by FDI under way, with commercial presence becoming the preferred mode of international expansion. But this could be a matter of the differential growth rates of different classes of service (more rapid growth of more FDI-oriented services), rather than a trend toward FDI within services sectors.

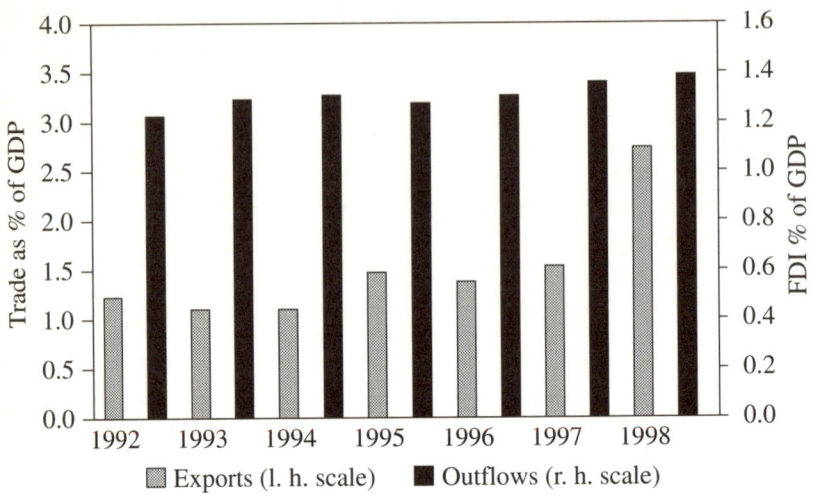

*Source:*    NewCronos; NEI calculations.

*Figure 3.1    Intra-EU trade and FDI flows in services relative to GDP*

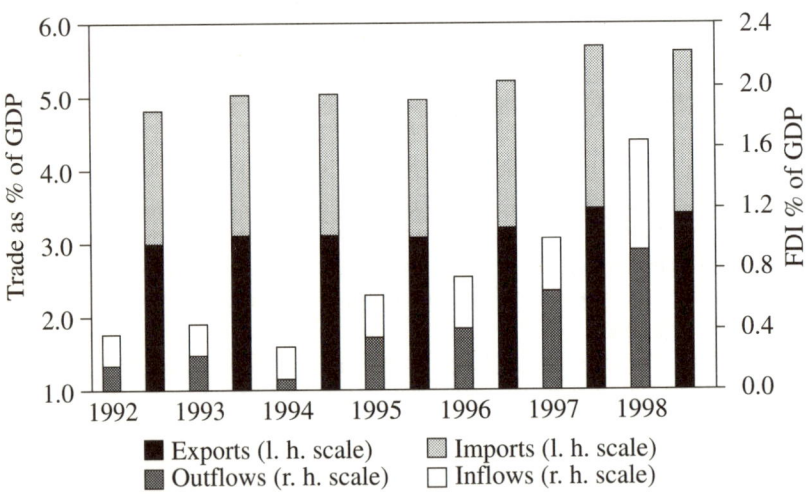

*Source:*    NewCronos; NEI calculations.

*Figure 3.2    Extra-EU trade and FDI flows in services relative to GDP*

Turning to the relative importance of European integration as compared to integration of Europe in worldwide service markets, analysis of aggregate trade and FDI flows data provides a mixed message. As shown above, trade in services is much less important than trade in goods, despite the dominance of services in UK employment and output. At less than 20 per cent of total trade, the share of services in total intra-EU trade changed little during the 1990s (see Table 3.9), despite the continuing trend towards a service economy. Similarly, the share of intra-EU imports in total service imports and the share of intra-EU exports in total services exports are largely unchanged; if anything, there has been a slight fall in the relative importance of intra-EU trade as recorded in these data. Thus, although trade in services is growing, the internal market – rather surprisingly – has not resulted in faster growth in trade within the EU than of EU trade with the rest of the world. Is this because the dynamics driving services trade with other regions are stronger, or because other modes of internationalization are more prominent within the EU?

FDI data do tell a different story. The share of services in total intra-EU FDI inflows rose considerably, from 63·5 per cent in 1992 to 74·7 per cent

*Table 3.9* The share of services in intra-EU trade and FDI, and the intra-EU share in total services trade and FDI (per cent)[a]

|  | 1992 | 1993 | 1994 | 1995 | 1996 | 1997 | 1998 |
|---|---|---|---|---|---|---|---|
| Services as a share of total intra-EU imports | 19.6 | 21.5 | 20.6 | 19.1 | 19.3 | 19.2 | 18.9 |
| Intra-EU imports as a share of total services imports | 57.7 | 57.7 | 57.9 | 58.7 | 57.3 | 55.6 | 55.8 |
| Intra-EU exports as a share of total services exports | 54.8 | 55.5 | 55.7 | 55.8 | 54.9 | 53.0 | 54.2 |
| Services as a share of total intra-EU FDI inflows | 63.5 | 58.9 | 64.2 | 68.7 | 72.2 | 71.8 | 74.7 |
| Intra-EU FDI inflows as a share of total services inflows | 62.7 | 61.7 | 63.1 | 61.9 | 57.7 | 63.9 | 58.3 |
| Intra-EU FDI outflows as a share of total services outflows | 75.9 | 66.7 | 87.0 | 63.6 | 58.3 | 48.7 | 54.5 |

*Note:* [a] Trade data exclude royalties and licence fees, and government services in the definition of services.

*Source:* NewCronos; NEI calculations.

in 1998. Integration in services within the EU may, then, be increasing more rapidly than in other sectors. The FDI flows provide an indicator of the establishment and strengthening of commercial presence in foreign markets across the EU, and we have seen that this is potentially a more important component of integration of services sectors than for goods-producing sectors.

At the same time, comparing intra-and extra-EU flows indicates an increasingly outward orientation of EU FDI. The share of intra-EU inflows has remained relatively stable, at around 60 per cent of total FDI inflows for services. But the picture is different for outflows. The share of intra-EU outflows in total outflows for services has declined quite dramat-ically, from 75 per cent in 1992 to only 55 per cent in 1998. Thus, although integration of EU service markets through FDI is growing rapidly, there is an increasing emphasis on investment outside the EU by European inves-tors.

The geographical composition of EU services transactions with the rest of the world is shown in Table 3.10. As can be seen, the USA is the single most important partner, notably with respect to inward foreign direct investment, both for mergers and acquisitions (that is operations where an EU firm is the target) and for flows and stocks of FDI assets (that is, EU liabilities). The composition of inward transactions (including FDI stocks) indicates that developed country regions account for the vast majority of international activities of non-EU service providers within the EU. By con-trast, the higher shares of other country regions for outward indicators point to a more diversified composition of international activities of EU service providers. This points to an increasingly 'global' dimension to the outward development of EU service sectors, which can largely be explained by the growing importance of EU relations with Central and Eastern European countries and, also, with other industrializing regions (for example, Latin America and Southeast Asia).

What about the sectoral composition of services trade and investment? Table 3.11 analyses the sectoral composition of extra-EU trade and FDI. It can be seen that travel and transport account for almost 60 per cent of trade. In contrast, finance and insurance account for 40 per cent of services FDI flows and stocks (perhaps coincidentally, the different proportions in the cases of flows and stocks add to very similar figures). Business services show very different patterns, too: computer services and R&D are more predominant in trade (though they are a small proportion of this), while the larger 'other business activities' category is more prominent in FDI. Services internationalization evidently proceeds at different paces and in different ways according to what types of services we are considering.

There are data on foreign affiliates trade in services (displayed in Table

*Table 3.10  Geographical composition of extra-EU transactions in service*

| | Trade[a] (1998) | | FDI flows (1998) | | FDI stocks (1997) | | M&A numbers (1996–8) | |
|---|---|---|---|---|---|---|---|---|
| | Exports | Imports | Outflows | Inflows | Assets | Liabilities | EU bidder | EU target |
| Value (ECU bn)[b] | 220.3 | 202.0 | 69.5 | 54.8 | 305.4 | 277.0 | 1389 | 1792 |
| Shares (%) | | | | | | | | |
| USA | 32.5 | 32.5 | 29.1 | 59.9 | 45.7 | 45.2 | 34.0 | 55.9 |
| Canada | 2.3 | 2.4 | 2.1 | 0.7 | 3.9 | 1.7 | 3.9 | 3.3 |
| Japan | 5.9 | 3.7 | 0.2 | 3.1 | 1.6 | 10.1 | 1.2 | 2.7 |
| EFTA[c] | 15.8 | 14.5 | 19.8 | 21.6 | 11.4 | 25.6 | 11.7 | 11.1 |
| Other countries | 43.5 | 47.0 | 48.7 | 14.7 | 37.4 | 17.5 | 49.2 | 27.0 |

*Notes:*
[a] Excluding royalties and licence fees and government services (not included elsewhere).
[b] Number of operations for M&A.
[c] Norway and Switzerland only for M&A.
Owing to rounding, some columns may not add up to 100.

*Source:*  NewCronos, AMDATA and Commission Services; NEI calculations.

*Table 3.11    Sector composition of extra-EU trade and FDI for services (per cent)*

| Sector | Trade service 1996–8 | FDI output flows 1996–8 | FDI output stocks 1997 |
|---|---|---|---|
| Trade and repairs | — | 14.6 | 17.3 |
| Merchanting and other trade-related services | 6.1 | — | — |
| Travel | 29.0 | — | — |
| Hotels and restaurants | — | 0.0 | 2.2 |
| Transport | 29.0 | 0.8 | 1.4 |
| Communication services | 2.1 | 7.0 | 1.3 |
| Insurance | 2.4 | 3.7 | 12.4 |
| Financial | 4.2 | 36.7 | 28.6 |
| Real estate | — | 0.3 | 5.7 |
| Construction services | 4.7 | — | — |
| Operational leasing | 0.9 | — | — |
| Computer and information | 2.1 | 2.2 | 0.9 |
| Research and development | 2.3 | 0.6 | 0.3 |
| Other business activities | 15.4 | 21.0 | 25.8 |
| *Of which:* | | | |
|   Legal, accounting, management etc. | 2.9 | — | — |
|   Advertising, market research etc. | 1.6 | — | — |
|   Architectural, engineering and other technical | 3.7 | — | — |
|   Agricultural, mining, on-site processing | 0.2 | — | — |
|   Other miscellaneous business services | 3.8 | — | — |
|   Services between affiliated enterprises n.i.e. | 3.3 | — | — |
| Other services | 1.8 | 2.6 | 3.4 |
| Total | 100.0 | 89.5 | 99.3 |

*Source:* NewCronos; NEI calculations.

3.12) that show a high presence of non-national firms in 'knowledge-intensive' service sectors (computer and related activities; advertising; architectural, engineering and technical testing and analysis; and so on). We can compare those to data on trade in services, which suggest that in such services trade appears less important than production (turnover) of foreign affiliates. These services, believed to be strategically important as vectors of technology transfer and the development and application of business knowledge, are presumably more dependent on face-to-face contact than on standardized information flows, and on the sorts of inter-

*Table 3.12  Comparison of share of non-nationally owned enterprises in turnover and ratio of trade to turnover (per cent)*

| Sector | Turnover of non-national enterprises in total turnover (1994, 1995) | Exports relative to total turnover (1996) |
| --- | --- | --- |
| Computer and related services | 21.4 | 6.4 |
| Architectural and engineering activities, technical testing and analysis | 20.6 | 8.8 |
| Advertising, market research and public opinion polling | 21.8 | 5.1 |
| Miscellaneous business services | 16.1 | 5.9 |

*Source:* NewCronos; NEI calculations.

action that demand local presence rather than those that can be achieved through fleeting visits.

## CONCLUSION

Changes in the regulatory environment and in new information and communication technologies underlie the current internationalization of services, which is taking place both through conventional exports and through investment.

Despite the prospects of greater services internationalization, this has so far only partly been realized, at least as documented by the available statistics reviewed here. While services dominate the economies of the EU, trade in services in the EU accounts for only about 20 per cent of total trade, and trade in services is more important for the EU than for other Triad members. Investment, however, is another story. FDI in services accounts for about 55 per cent of total FDI flows and stocks. It may well be that we should not expect to see a 'catching up' of services exports so much as a continuing substantial role of services in investment flows and in mergers and acquisitions.

Trade is more important for smaller member states, who have the highest trade to GDP ratios. This suggests that there might be stronger motivations for firms to trade (and undertake FDI) and for governments to create incentives to encourage these interactions. (The argument may be that, since economies of scale are important for services as a whole, smaller countries with correspondingly small domestic markets tend to be more 'open' to services internationalization. Other arguments may concern the importance of technology transfer and learning from good international practice).

Service sectors in the larger economies are more outwardly (that is extra-EU) oriented. This, most probably, is a reflection of two factors. First, the larger member states are home to the EU's largest multinational service firms, which have the greatest worldwide reach. In other words, the domestic (home) market size may be influential in determining the extent of the geographical projection of service sectors. Second, the size of the domestic market within member states may render it attractive as a location for foreign firm affiliates whether they are acquired through mergers and acquisitions or result from greenfield investments. This seems to be so especially for firms from outside the EU.

Should further analysis confirm that larger economies are more outwardly oriented, this could be an important determinant of the long-run implications of the internal market. To suggest one such implication: if European service markets become integrated and location within the EU becomes less important for defining a firm's 'home' market, the ability of firms from smaller member states to increase their geographical reach may be enhanced. This could lead to a different pattern of players in international services in the years to come.

## NOTES

1 This chapter draws upon work undertaken by Paul Baker as part of a background paper to the European Competitiveness Report 2000 prepared for the European Commission, Directorate General Enterprise. The research assistance of Nora Plaisier and Viera Spanikova is gratefully acknowledged.
2 The court indicated in a number of judgments the activities subject to the free flow of services. See, for example, case no. 36/74, Walrave-Kochl; case no. 13/76, Dona; case no. 33/74, van Binsbergen.
3 The possibility of immediate electronic delivery can make the possibility of 24-hour trading more relevant for services than for goods.
4 Hipp *et al.* (2000), found surprisingly high standardization reported in many German services firms.
5 For example, travel and transport account for around 60 per cent of extra-EU trade in services.
6 After allowing for the importance of transport and travel in total cross-border trade in services.
7 Related party trade refers to trade between a US-owned or EU-owned parent firm and its US or EU affiliate.
8 The definition of services used here covers royalties and licence fees, and other private services (education, financial services, insurance, telecommunications, business and technical services and other services).
9 Balance of payments (BOP) statistics cover presence of (foreign) natural persons if they are resident for less than one year, longer visits mean they are counted as residents.

# REFERENCES

Deardorff, A.V. (1985), 'Comparative advantage and international trade and investment in services', in R.M. Stern (ed), *Trade and investment in services: Canadian/US perspectives*, Toronto: Ontario Economic Council.

Dunning, J.H. (1989), 'Multinational enterprises and the growth of services: some conceptual and theoretical issues', *Service Industries Journal*, **9** (1), 5–39.

Fontagné, L. (1999), 'Foreign direct investment and international trade: complements or substitutes', STI Working Paper DSTI/DOC (99) 3, OECD, Paris.

François, J.T. (1990), 'Trade in producer services and returns due to specialization under monopolistic competition', *Canadian Journal of Economics*, **23** (1), 109–24.

Goedegbuurte, R.V. (1996), 'Global statistics', *Netherlands Official Statistics*, 11, CBS, Heerlen.

Hatzichronoglou, T. (1996), 'Globalization and competitiveness: relevant indicators', STI Working Paper 1996/5, OECD, Paris.

Hindley, B. and A. Smith (1984), 'Comparative advantage and trade in services', *World Economy*, **7** (4), 369–90.

Hipp, C., B. Tether and I. Miles (2000), 'The Incidence and Effects of Innovation in Services: Evidence from Germany', *International Journal of Innovation Management*, **4** (4), 417–54.

Karsenty, G. (1999), 'Just how big are the stakes?: an assessment of trade in services by mode of supply', World Trade Organisation, April.

Peneder, M. (2000), 'External Services, Structural Change and Industrial Performance', WIFO background report for the Competitiveness Report 2000, DG-Enterprise.

Rubalcaba-Bermejo, L. (1999), *Business Services in European Industry*, European Communities, Brussels, Luxembourg.

Sampson, G.P. and R.H. Snape (1985), 'Identifying the issues in trade in services', *World Economy*, **8** (2), 171–81.

Sapir, A. and C. Winter (1994), 'Services trade', in D. Greenaway and L. Alan Winters (eds), *Surveys in International Trade*, Oxford: Blackwell.

Soete, L. and M. Miozzo (1989), 'Trade and development in services: A technological perspective', Working Paper No 89–031, MERIT, Maastricht.

Stevens, G. and R.E. Lipsey (1992), 'Interactions between domestic and foreign investment', *Journal of International Money and Finance*, **11** (1), 40–62.

UNCTAD (1997), *World Investment Report*, Geneva: UNCTAD.

Vandermerwe, S. and M. Chadwick (1989), 'The internationalization of services', *Services Industry Journal*, **9** (1), 79–83.

WTO (1995), 'GATS and statistics on trade in services', S/C/W/5, July.

WTO (1998a), 'Working Group on the Relationship between Trade and Investment: Communication from the United States', WT/WGTI/W/14.

WTO (1998b), 'Availability of statistics on foreign investment and on the activities of foreign affiliates', WT/WGTI/W/29, February.

# APPENDIX 1: SERVICES: MODES OF SUPPLY AND STATISTICAL DATA SOURCES

It is generally true to say that available statistical sources do not clearly identify or measure precisely the different modes of services supply. This appendix provides a description of some of the salient features of the data used in the chapter.

## Conventional Trade in Services (BOP) Data

With regard to mode one and mode two, cross-border supply and consumption abroad are covered by the current account of the balance of payments (BOP) statistics; for consumption abroad these are mainly covered by the BOP category of travel. Unfortunately, however, BOP statistics do not distinguish between cross border supply, consumption abroad and presence of natural persons (that is, mode four).

Moreover, with regard to the presence of natural persons, foreign natural persons are excluded from BOP statistics if they intend to stay for longer than one year or if effectively they stay for such a period. In either case they are counted as residents and the output they generate and that is sold in the host market is not recorded as trade in BOP statistics.

More generally, a basic problem arises with BOP statistics because the intangibility of services precludes their registration at the frontier as happens in the case of goods. Consequently, transactions are estimated, using either foreign exchange records or surveys of establishments, or a combination of the two. Difficulties in ensuring the reliability of such information and its attribution to different forms of service transaction imply that a question mark must remain over the precision of BOP data for services transactions. Further complications arise because of definitional differences and coverage across countries, although in the current case the data presented in this chapter are drawn from harmonized Eurostat sources. Given the aforementioned statistical problems associated with services trade data, BOP data should be treated as only rough proxies for services trade under mode one and mode two.

In principle, BOP data under the heading of royalties and licence fees, which includes franchise fees, could provide some indication of third party methods of obtaining commercial presence (mode three). Available data do not, however, make a distinction between types or sectors and so such analysis is not possible. For the analysis contained in this chapter, royalties and licence fees are not included in the definition of services trade as they do not relate to a specific services activity.

**Foreign Direct Investment Data**

With regard to mode three, commercial presence, a distinction may be made between statistics on foreign direct investment (FDI), those on mergers and acquisitions (M&A) and those on the activities of foreign affiliates (FATS).

Foreign direct investment is a capital flow that is a key element in the capital and financial account of the balance of payments. Related to it is the income on foreign direct investment recorded in the current account and the international investment position (that is, FDI stocks). The generally accepted international agreement on the definition of FDI covers capital flows from a 'direct investor' resident in one country to a 'direct investment enterprise' resident in another country, reflecting the objective of obtaining a lasting interest of the direct investor in the direct investment enterprise. Recorded FDI flows cover both the initial capital transaction and all subsequent transactions. For practical purposes, a cut-off is established such that FDI transactions occur for BOP purposes when the direct investor obtains 10 per cent or more of the ordinary shares or voting power in the direct investment enterprise.

Statistics on FDI are collected in the context of compiling BOP statistics, typically by the resident central bank. Normally a breakdown may be made between equity, reinvested earnings and inter-firm debt transactions. With regard to reinvested earnings, unlike capital flows or distribution of dividends (FDI income), there is no accompanying foreign exchange transaction, thus making this the hardest category of FDI flows to estimate. In this chapter, harmonized data based on a joint Eurostat/OECD questionnaire and published by Eurostat have been used and, in the light of the previous comment, FDI flows are measured as the sum of direct equity investment and other capital only (that is, excluding reinvested earnings).

It should also be noted that FDI statistics capture only very partially the total investments of foreign affiliates. Multinational firms raise capital from a variety of sources both at home and abroad (for example, commercial bank loans, local and international equity markets, public organizations and reinvested profits). Taking these various forms into consideration, UNCTAD (1997) estimates suggest that FDI flows represent only a quarter of investment in foreign affiliates.

**Foreign Affiliates Trade Data**

Foreign affiliates trade statistics (FATS) relate to the activities of foreign firms and may cover a wide variety of different indicators on the domestic and foreign operations of multinational firms (turnover, employment, value added, investment and so on). Thus, despite their name, the data cover variables that are non-trade related. An important distinction between FDI data from the BOP and FATS relates to the population of enterprises covered. For FDI statistics, a minimum of 10 per cent of foreign ownership is used as a cut-off point. By contrast, for the collection of FATS data, a minimum of 50 per cent is used. Accordingly, the statistical population for FATS statistics represents only a subset of the whole FDI population.

The data used in this chapter are drawn from pilot studies initiated as part of a joint Eurostat/OECD project. They cover inward FATS only, that is to say, they are concerned with the activities of foreign affiliates in the domestic economy of the reporting country, but do not cover outward FATS (that is, activities of domestic enterprises of the country studied in foreign markets).

# 4. Internationalization of services: are the modes changing?

## Zbigniew Zimny and Padma Mallampally

### INTRODUCTION[1]

One of the striking features of globalization in recent decades is the shift in the modes of internationalization of economic activity towards transnational corporation (TNC)-related modes and transactions. As a result, according to estimates by the *World Investment Report 2001*, in 2000, the value of goods and services sold by foreign affiliates of TNCs was more than twice that of exports: $16 trillion versus $7 trillion (UNCTAD, 2001, p. 10). Foreign direct investment (FDI) and trade have become increasingly interconnected,[2] and the role of TNC-related trade in total trade is significant. For example, in 2000, exports of foreign affiliates from host countries alone amounted to 46 per cent of world exports. In the United States – the largest home and host country for foreign direct investment (FDI) in the world – TNCs, national and foreign, account for three quarters of total exports. Over a third of the total is intra-firm exports (UNCTAD, 1999, p.232). Although these data cover both goods and services, goods dominate and, consequently, these relationships are characteristic of the internationalization of goods rather than services.

The objective of this chapter is to examine the role of TNCs in international transactions in services, whether this role has changed, and whether the modes of internationalization in services are becoming similar to those characteristic of internationalization in goods. The chapter is organized in two parts. First, we review briefly the reasons why the modes of delivery of, or transactions in, services should be changing, and in what direction. The focus is on the impact of technology and its implications for the internationalization of services firms or activities and the organizational strategies of service and goods TNCs as regards international production of services. Secondly, we look at what the data show as regards modes of services delivery and transactions, focusing on trade and FDI – the latter, specifically, as reflected in the sales of foreign affiliates. We begin by examining the relative importance of trade and FDI. We then look at the role of TNCs in trade itself, examining intra-firm trade as an indicator of integrated international production of services. Our interest is mainly in shifts in the modes of delivery of, and transactions in, services that can be delivered by both modes, caused, in particular, by technology and its implications for the ways in which services production is organized internationally.

## REASONS FOR CHANGING MODES

### The Impact of Technology

There are a number of ways in which the technological developments that have occurred in recent decades might be expected to have influenced the internationalization of service industries and the modes of international delivery of services. First, advances in telecommunication and data technologies have dramatically increased the transportability of some services, especially of information-intensive services, creating scope for cross-border trade at arm's length. Many of the services affected are intermediate rather than final products (UNCTAD, 1994, pp.53–7; 1996, p.105; Miozzo and Soete, 2001, pp.164–5).

Secondly, new product and process technologies, hard and soft, for services production have enabled the innovating firms to acquire ownership-specific (O) competitive advantages. According to the ownership, location and internalization (OLI) paradigm, O-advantages are a necessary (though not sufficient) condition for firms to expand abroad through FDI; FDI will follow if there are advantages to firms of internalizing their transactions, and may also follow if physical interaction with or proximity to customers is necessary. Once they expand, firms acquire experience in 'multinationality', enhancing further, through their expertise in organizing and managing international production, their O-advantages (Dunning, 1989).

Thirdly, new telecommunication and data technologies have facilitated management and coordination of internationally dispersed activities and reduced the costs of management and coordination, easing FDI.

Fourthly, technological progress in service industries (and the increased competition as well as development potential it implies) has prompted governments to open up to FDI in many industries, such as telecommunications, power generation and distribution, air transport and banking which were considered strategic and thus hitherto reserved for domestic control and ownership. Opening to FDI, including through privatization programmes enhances the L-advantages (in the language of the OLI paradigm) of many potential host countries, creating an inducement for FDI in services.

Through the interaction of these influences, technologies can affect the choice of mode of services delivery and the strategies of TNCs with respect to international production in services. Specifically, increased tradability of services side-by-side with greater incentives to establish production operations abroad as a result of newly acquired ownership and locational advantages creates the prospect that the international production strategies of firms in services will become more similar to those pursued by TNCs in manufacturing and other goods production, reconfiguring the ways in which services are produced and exchanged internationally.

## Are TNC Strategies in Services Shifting towards Integrated International Production?

Over the past few decades, the strategies of TNCs with respect to the organization of international production in goods – involving decisions about the international location of different activities and the degree of integration among the various entities that fall under the common governance of the firm – have evolved and broadened from largely 'stand-alone' strategies to integrated international production strategies, simple or complex.[3] Under stand-alone strategies, a foreign affiliate acts largely as an independent entity within host economies and is assigned responsibility for most of the functional activities along the value chain of a product. Under integrated strategies, a foreign affiliate is responsible for activities along a portion of the value chain of a TNC and its output is integrated with that of the parent or of other affiliates. Organizational structures have also changed, accordingly, from the establishment of independent, almost 'fully equipped' affiliates on a multidomestic basis (linked to the parent firm primarily through ownership and technology), with intra-firm flows limited to those of equipment and services from the parent to each affiliate, to regional and global networks of interdependent firms within TNC systems, based on specialization between parent and affiliates as well as among affiliates. While the range of possible strategies and structures has grown over time as TNCs have responded differently to major changes in the international economic environment and technological progress, there is a trend among TNCs in many industries to adopt strategies and structures that involve closer integration of their functional activities, whether by giving primary responsibility for a function to a foreign affiliate rather than the parent or by sharing the performance of functions among several foreign affiliates or the affiliates and parent.

Simple integration strategies involve the establishment of foreign affiliates in which a limited number of activities along the value chain are performed, and linked to those done elsewhere within the production system of a TNC. These strategies aim primarily at exploiting the locational advantages of host countries – typically, natural resources or low-cost labour – to gain access to resources and/or to reduce the cost of production of a specific portion of the value chain. FDI in the extraction or cultivation of primary commodities for export to the parent firm or other affiliates for further processing and/or sale in the home country and other countries has long been driven by such strategies. In the more recent past, the pursuit of such strategies has been an important force behind FDI in industries such as clothing, toys, semiconductors and other electronic products.

While simple integration strategies focus on establishing a limited number of linkages within TNCs' production systems in order to take advantage of opportunities for specialization and cost minimization in specific activities, complex integration strategies involve the location of various functional activities along the entire length of the value chain wherever they can be done best in terms of fulfilling a firm's overall objectives (UNCTAD, 1994, p.121). These strategies are based on a firm's ability to shift production or supply to wherever it is most profitable: any affiliate operating anywhere may perform functions for the firm as a whole and is judged in terms of its contribution to the entire value chain. There is substantial functional integration among the different processes and the different locations. Management and coordination are integrated, to some extent, across locations in all three strategies mentioned, generally (but not invariably) being provided by the parent firms; however, the complexity of integration of these functions across different units of a TNC system increases with the complexity of integration of other functions along the value chain.

The strategies and organizational structures of some TNCs in the automobile industry provide illustrations of integrated production strategies in manufacturing. Several major car manufacturers have adopted strategies that involve complex integration both across the functions performed by parent and affiliates and between them and those of other firms with which the former have non-equity linkages, and across a wide geographical area. Prominent examples include the following:

1.  Ford's network in Europe. After decades of operating on a stand-alone, multidomestic basis throughout Europe, Ford Motor Company reorganized its production to integrate the operations of its European affiliates in the late 1960s, taking advantage of falling trade barriers within Europe to movements of components and final products as a result of the establishment of the European Economic Community (now the European Union). Product development, component manufacturing

and final assembly were spread out among Ford's various affiliates in Europe during the 1970s and 1980s to take advantage of specialization and scale economies (UNCTAD, 1993, pp.147–8).[4] Subcontractors and suppliers to Ford were also linked across borders.

2. Toyota's network in Asia. Responding to the subregional industrial cooperation policies of the ASEAN, Toyota Motor Company established, in the mid-1990s, a network of affiliates in ASEAN member countries for the manufacture and supply of parts to its vehicle-manufacturing plants in Southeast Asia and Japan.[5] Coordination of intra-firm exports from these locations was performed by a Toyota affiliate in Singapore (UNCTAD, 1996, p. 100). More recently, the functional and geographic scope of Toyota's integrated production within the region has expanded to include more countries and more activities (UNCTAD, 2001, p.87).

3. Honda's interregional network. Beginning in the 1960s with the establishment of assembly affiliates that relied upon supplies from Japan, Honda Motor Company's motorcycle operations in Europe had, by the mid-1980s, become closely integrated and linked to its activities in Brazil, Japan and the USA. Two of its four affiliates in Europe began to produce engines and a third, other parts and components, in addition to assembling the vehicles. Intra-firm supply links were established within Europe as well as with Honda's affiliates in Brazil and the USA and Honda's plants in Japan (UNCTAD, 1996, p.102).

Although complex integration of the kind illustrated above, with numerous links between units of TNC production systems dispersed regionally or globally, is best exemplified by large automobile firms, it can also be found today in other manufacturing industries, such as computer electronics.

In the past, the scope for TNCs to pursue integrated strategies in services has been limited, the main (but not the only) reason being that, with the exception of services that could be embodied in a physical form (for example, books or audiotapes that embody final services such as education and entertainment and documents that embody intermediate services such as engineering designs and legal advice), services were not transportable and had to be consumed in simultaneous interaction between producers and consumers. International transactions therefore required the temporary movement of a producer to a consumer or vice versa, or the establishment of production operations abroad. Firms could rarely split up their value chain and relocate, for instance, labour- or human capital-intensive service components of the chain to countries where they could be produced more cheaply or with more beneficial skill/cost ratios. In host countries, including host developing countries, they therefore had to reproduce factor proportions similar to those used at home. This meant that there were fewer opportunities for exploiting factor endowments through intra-firm trade; at the same time, there was greater developmental impact of service TNCs, as compared with manufacturing ones, in terms of transferring soft technologies, upgrading skills and offering higher wages in host countries (Kravis and Lipsey, 1988; UNCTC, 1989a, p.27, 1989b, pp.111–12).

The impact of the technological changes of recent decades should have been to give firms a greater choice of the modes of internationalization of service activities and to have enabled TNCs to pursue integrated strategies where profitable, resulting, potentially, in increased intra-firm trade in services. The latter applies not only to services TNCs but also to goods TNCs that are also engaged in producing services, either the entire value-added chain of final services or intermediate services that are a part of their value-added chain for goods.

Indeed, anecdotal evidence shows that integrated international production strategies in services are not only possible but actively pursued by firms. There is a variety of cases to illustrate TNCs in various industries locating one or more functional activities along the value chain of services in their affiliates abroad, and integrating them with activities elsewhere within their production systems:[6]

1.  Insurance: New York Life, since the late 1980s, has conducted part of its insurance payments processing in its affiliate in Ireland (UNCTAD, 1996, p.107). Customers' claims were sent overnight by air to the affiliate for processing, and returned after processing by dedicated telecommunication line to the firm's data-processing centre in New Jersey, where checks or responses were printed out and mailed to the clients. The motivation behind moving a part of the process to Ireland was to reduce labour costs and overcome difficulties in finding enough skilled personnel at home to process insurance claims.

2.  Airlines: Swissair transferred its revenue accounting (the calculation of amounts earned from and owed to other airlines by the carrier, on the basis of flight coupons collected) to an affiliate that it established (with 75 per cent ownership) in India in 1993 (UNCTAD, 1993, p.122). Flight coupons were flown to the affiliate, credits and liabilities were processed there and the results transferred to a central processing unit in Switzerland for informing each branch that originally sold the tickets. In addition, the Indian affiliate also processed records of Swissair's inter-airline transfers and transactions and, over time, performed other airline-related services (including settlement of queries in different administrative functions). American Airlines has similarly established an affiliate in Barbados for processing its accounts material

and ticket coupons (UNCTAD, 1996, p.107). The affiliate enters details of hundreds of thousands of American Airline tickets daily into a computer and the data are returned by satellite to the firm's data centre in the USA.

3.  Telecommunications: a unit of Telefonica (Spain) has built, along with outsourcers, call centres near Rabat and Tangiers at which more than 2000 Moroccans are employed to talk to Spanish and French consumers.

4.  Banking and financial services: Hong Kong and Shanghai Bank has established an affiliate in India for handling its back-office work. American Express has located finance and customer services in its affiliate in India.

5.  Computer and information-technology services: Cisco Systems, considered the world leader in networking for the Internet, has established, in addition to sales offices in all major markets, global software R&D labs, including one in Bangalore, India. It has announced plans for building a corporate campus in Amsterdam where almost 5000 people are expected to be employed – a centralized fulfilment centre servicing Europe, the Middle East and Africa. Cisco has also entered into a collaboration with Singapore's Advanced Research and Education Network for long-term R&D for next generation high-speed networks and Internet technologies.[7] Catalytic Software, a USA-based firm, maintains a small marketing and training force in its home-based operations and is planning to develop software mainly in India where it is developing a township near Hyderabad for its employees.[8] Intel, the world's largest computer chip manufacturer, has set up an R&D laboratory jointly with India's Infosys Technologies in Bangalore. Intel will provide the necessary hardware, software and technical information to enable Infosys to develop the software applications built on Intel's platforms.[9]

6.  Diversified services: General Electric Corp. has located accounting, claims processing, customer service, credit evaluation and research functions in its GE Capital unit in India. The firm employed 6000 people in India in 2000 to provide these services for GE operations around the world.

The examples cited above suggest that integrated international production strategies are being pursued in services in which one or more intermediate products are now tradable by electronic means and can be delivered without proximity between the buyer and the producer. However, most of them point to simple integration, based on FDI by TNCs seeking lower-cost human resources or efficiency and leading to increased intra-firm imports to the home country or a limited number of other destinations. Complex strategies that involve intra-firm trade among several units of a TNC network, including among affiliates regionally (for example, in Western Europe or NAFTA), or globally seem to be less common. Nevertheless, it seems plausible that TNCs in scale-intensive, information technology-based services may well pursue complex integration strategies comparable to those witnessed in the automobile industry.[10]

Integrated international production within TNC systems is not, of course, the only way in which firms can take advantage of greater choice of modes of delivery of services to exploit cost and efficiency advantages in different locations. Outsourcing to independent foreign producers of services is another option. Within economies, 'business process outsourcing', including outsourcing of some of the services mentioned in the examples above, has been growing in importance in recent years. The market for outsourced services is expected to grow rapidly – according to one source, from $119 billion in 2000 to $234 billion in 2005.[11] Outsourced services have expanded in recent years to include, in addition to professional services like advertising and business consulting (management, engineering and legal services), specific processes related to conducting a business, many of them considered 'back-office' functions when conducted within firms for which they are not the main output. These include services such as human resource management; payrolls; buildings management; insurance claims processing; information-technology services; finance and accounting services; premiums processing and claims management; credit cards processing and payment services; logistics and procurement. Growing experience with outsourcing a limited range of services, such as those related to information technology, is expected to give more firms the confidence to outsource a wider range of services and the skill intensity of outsourced services is expected to increase over time.[12] Externalization of the production of 'business-process' services such as those mentioned above, most of which are inputs for manufacturing as well as services firms, allows firms to lower costs because of the economies of scale, scope and skill that providers of services to a number of firms can enjoy. At the same time, the costs of interaction and management involved in outsourcing are declining, as a result of lower communication costs, standardization of web-based tools, and increasing automation of their own data services by client firms.

The trend towards outsourcing described above suggests several possibilities as regards internationalization strategies of firms with respect to such services. First, competitive firms specializing in the provision of outsourced business services may locate some of their functions in foreign affiliates that they establish in order to take advantage of internalization as well as locational advantages, the latter especially in the areas of human resource cost and skills, and integrate them through intra-firm trade with those produced at home and in other foreign affiliates for delivery to customers. Secondly, large firms that use such services, especially firms that are already TNCs, may choose to internalize the production of a service but resort to FDI in one or more functional activities in the service value chain to take advantage of cost or efficiency differences across locations while reaping economies of scale and scope at the same time. That, too, would involve integrated international production. Finally, some firms – whether providers or users – may outsource some services to foreign providers through contractual arrangements of various kinds. In some services, such as those related to information technology, such outsourcing is already evident.[13] Depending on the nature of the service involved, outsourcing can involve contractual relationships of various kinds between the service provider and the customer, allowing for control over quality by the provider and customization to suit the specific needs of individual customers. For example, EDS, a leader in the global information technology services industry,[14] with around 100 000 employees in over 40 countries, has outsourcing and cosourcing contracts that involve, among other things, the transfer of clients' staff to EDS (Hood and Peters, 2000, p.93).

To sum up, the likely impact of technological changes on the relative roles of trade and FDI in international transactions in services is not straightforward. The need for customer physical proximity is greater in some services than others and, hence, telecommunication and data technologies have not enhanced the tradability of all services to the same extent.[15] Also, technological advances enhancing firms' ownership advantages and potential for FDI have not affected all service industries equally.[16] Thus some of the changes are likely to have stimulated trade by increasing the tradability of some services; some, FDI (by enhancing O and L advantages in some non-tradable services) and some others, both FDI and (intra-firm) trade (by enhancing O and L advantages in services with enhanced tradability). The most evident change should concern intra-firm trade. Even in the case of such non-tradable services as hotel services, retail trade or power generation, expansion of FDI networks could lead to intra-firm trade in information-related, management and coordination services.

## WHAT DO THE DATA SHOW?

We use US data, first comparing exports with sales of foreign affiliates of US TNCs abroad and imports with sales of foreign affiliates of non-US TNCs in the USA to examine the relationship between trade, defined as transactions between residents and non-residents,[17] and FDI as modes of international delivery of services and see whether the forces discussed above have changed this relationship. Secondly, we examine trade and sales data for evidence on the extent of, and trends in, international production of services by TNC, including evidence on integrated international production.

The USA is the only country for which detailed data are available on the outbound and inbound sales of services by foreign affiliates and intra-firm transactions in services, in addition to arm's-length cross-border transactions, permitting a comparison of the modes of delivery on a time-series basis. The USA is also a very instructive country to look at to examine the role of TNCs in international transactions, including trade. It was the first country to generate services TNCs on a large scale across many service industries. It is still the source of the largest amount of outward FDI in services among countries. Moreover, when services TNCs from other countries emerged and expanded rapidly in the second half of the 1980s and during the 1990s, much of their expansion took place in the US market. As a result, the USA became the largest host country for FDI in services (Sauvant and Zimny, 1987, pp.30–2). So modes of delivery of services to the US market reflect strategies of service TNCs from other countries. We look at changes over time, as far back as the data available permit, in most cases since 1986–7.

### FDI or Trade, or Both?

During the 1980s, just prior to the GATS negotiations within the Uruguay Round of multilateral trade negotiations, the common perception was that FDI was the dominant mode of services delivery to foreign markets (Sauvant, 1990, p.115). Many services were not transportable and hence non-tradable at arm's length, which led in many cases to the need to establish foreign affiliates to produce and sell them abroad. (In some cases, this could be done through the temporary movement of individual consumers or producers – such sales are classified as trade for balance-of-payments purposes). This perception was reinforced by the visible role of US service TNCs across many service industries,[18] and the fact that the USA was a key proponent of the inclusion of services in the multilateral trading system (Miozzo and Soete, 2001, pp.178–9). But the dominance of FDI is not supported by the data for those years. Even in the USA, the principal home country for service TNCs in the aggregate, exports (broadly defined as transactions between residents and non-residents) were almost as important as a mode of delivery of services abroad as were sales of services through majority-owned foreign affiliates (MOFAs). In most years during the second half of the 1980s and the early 1990s, the ratio of sales of services by US TNCs' MOFAs to US exports of services exceeded one only slightly. Only in two years between 1986 and 1995 (1987 and 1988) was it higher than 1·20. For the inbound relationship illustrating the modes of delivery of non-US firms, the ratio was initially clearly below one, amounting to 0·86 in 1987. It should be noted that these ratios would be even lower if fees and royalties were included in trade data. In comparison, at that time international delivery of *goods* already took place predominantly via the FDI mode: in the second half of the 1980s, the ratio of sales of goods by US foreign affiliates abroad to US exports of goods ranged between 2·5 and 3, oscillating thereafter around 2·5 (Table 4.1).

Since the mid-1990s, there has been a noticeable increase in the ratios of foreign affiliate sales to trade in services, with the ratios for both outbound and inbound transactions consistently exceeding 1·5, heading towards the level of two and, thus, convergence with the ratios characteristic of the international delivery of goods. These changes suggest that the impact of increased arm's-length tradability of information-intensive services has thus far not shifted the balance of international transactions in services, taken as a whole, in favour of trade between partners (in this case, the USA and the rest of the world taken as a whole), as compared with FDI-related sales. On the contrary, the situation, in the most recent years, has been the opposite.

Considering that the data on foreign affiliate sales relate only to MOFAs and, hence, underestimate total affiliate sales,[19] and that, moreover, the value of transportation transactions comprises a particularly large fraction of total trade, this suggests that, overall, in services other than transportation, the factors favouring FDI in services (for example, ownership advantages coupled with internalization advantages for innovating firms, liberalization of TNC entry to major service industries) have had a stronger influence on the pattern of internationalization of services activities than have those increasing tradability (for example, lower cost and ease of electronic delivery). Moreover, given that, as shown below, a very high and growing proportion of sales of services by US foreign affiliates abroad are sales to local (host country) markets, the importance of presence and/or proximity for delivery to local markets seems to be a major factor determining the greater importance of FDI relative to trade. In this context, it is important to note that the pattern for goods (always tradable) shows an even greater reliance on FDI than on trade.

The importance of factors requiring presence and proximity becomes clear when the services for which disaggregated data are available are divided into three categories classified by modes of delivery: (a) services delivered mainly or only via trade (travel, passenger fares and other transportation); (b) services delivered mainly or only via FDI (such as trading

Table 4.1  Ratio of FDI sales to exports and imports, goods and services, United States, 1986–99

| | 1986 | 1987 | 1988 | 1989 | 1990 | 1991 | 1992 | 1993 | 1994 | 1995 | 1996 | 1997 | 1998 | 1999 |
|---|---|---|---|---|---|---|---|---|---|---|---|---|---|---|
| Goods: outward ratio[a] | 2.85 | 2.87 | 2.55 | 2.46 | 2.71 | 2.58 | 2.53 | 2.38 | 2.45 | 2.52 | 2.59 | 2.45 | 2.47 | — |
| Services: outward ratio[b] | 1.16 | 1.24 | 1.22 | 1.02 | 1.06 | 1.05 | 1.04 | 1.03 | 1.04 | 1.14 | 1.23 | 1.29 | 1.42 | 1.58[d] |
| Services: inward ratio[c] | — | 0.86 | 0.92 | 1.06 | 1.19 | — | 1.29 | 1.44 | — | 1.21 | 1.28 | 1.56[d] | 1.57 | 1.75 |

*Notes:*
[a]  Ratio of sales of goods by US MOFAs abroad to US exports of goods.
[b]  Ratio of sales of services by US MOFAs abroad to US exports of services.
[c]  Ratio of sales of services by affiliates in the USA of TNCs of countries other than the USA to US imports of services.
[d]  In 1997 for inward sales and in 1999 for outward sales there was a change in the industry classification system which resulted in the shift of sales of affiliates from goods to services. Under the previous system, sales of services were underestimated. However, changes in the classification system were not decisive in the shift of the ratio towards FDI sales. For the outward ratio it began before the change and for the inward ratio it continued after the change of system. In addition, the shift was not that big. For example, only half of the 18 per cent of the growth of sales of foreign affiliates of US TNCs in 1999 was due to the change of the system (US Department of Commerce, 2001, p.49).

*Source:*  *Survey of Current Business*, various issues.

*General note:* sales are total sales of majority-owned affiliates (MOFAs). Trade in services excludes royalities and fees. It includes travel, passenger fares, other transport and other private services. Insurance is on a gross basis (premiums not adjusted for losses) to make it comparable with sales which are gross sales. Sales of services exclude banking services for which data are not available. Trade in services includes services provided by banks.

and hotel services and power generation and distribution); and (c) services that can be delivered by both modes (business, professional, telecommunication and financial services). The third category includes services the tradability of which has increased, giving firms a greater choice between modes of delivery. It has been the fastest-growing category in terms of total international transactions, which increased almost fivefold during 1986–99 (Table 4.2). But sales of these services through MOFAs increased faster than exports, and much faster than sales of the second category of services delivered mainly or only via FDI. Inward delivery of services into the USA has undergone a similar change: for the category of services with increased choice of the modes of delivery, sales increased much faster than imports. This underlines the increased importance of FDI for services for which tradability theoretically increased: for example, financial, business and professional services. Apparently, tradability was not the most important among the many factors determining the mode of delivery. One should not, however, go as far as to interpret this as FDI replacing trade. Both exports and imports of services in the third category increased rapidly, much faster than those of services delivered mainly via trade, pointing to complementarity of trade and FDI, rather than their substitutability.

Data on international transactions at the aggregate level conceal, as might be expected, differing patterns with respect to affiliate sales and trade for different service industries belonging to the third category (Table 4.3). But with a few exceptions, the individual industry pattern is consistent with the general pattern observed earlier. First, in the late 1990s, for most industries in this category except for accounting, R&D and management services in inward transactions and health services in outward transactions sales by MOFAs exceed trade. Secondly, the increased tradability of services not only has not decreased the role of FDI as a mode of delivery but in most industries this role further increased, a notable exception being advertising, where the sales to trade ratio has decreased, but from a very high level.

Among the individual service industries, the industry with the highest ratio of foreign affiliate sales to trade on the outward side of US international transactions is the category of computer and data-processing services, followed by advertising. As for inward transactions, advertising, computer and data-processing services, and construction, engineering and mining show the highest foreign affiliate sales relative to imports. The increasing importance of foreign affiliate sales relative to exports in computer and data processing services is particularly striking, suggesting that presence in local markets is particularly important for competitiveness in this industry which technically should have been strongly affected by increased tradability of services. Although the sales/trade ratios declined in advertising, FDI remains the dominant mode of delivery in this industry,

*Table 4.2    Growth of international transactions in services by modes of delivery, United States, 1986–99 (millions of dollars and per cent)*

| Service category | 1986 | 1999 | Increase 1999 (per cent 1986 = 100) |
|---|---|---|---|
| | (millions of dollars) | | |
| A.  Outward transactions | | | |
| 1. Services delivered mainly via trade[a] | 41405 | 121432 | 293 |
| 2. Services delivered mainly via FDI[b] | 26655 | 85802 | 322 |
| 3. Services delivered via both modes: | | | |
|    a. Exports[c] | 28027 | 104126 | 372 |
|    b. Sales[d] | 46194 | 252607 | 547 |
|    Subtotal (a + b) | 74221 | 356733 | 481 |
| Total outward (1 + 2 + 3) | 142281 | 563967 | 396 |
| | 1987 | 1999 | Increase 1999 (per cent 1987 = 100) |
| | (millions of dollars) | | |
| B.  Inward transactions | | | |
| 1. Services delivered mainly via trade[a] | 55603 | 114319 | 206 |
| 2. Services delivered mainly via FDI[b] | 17245 | 87224 | 506 |
| 3. Services delivered via both modes: | | | |
|    a. Imports[c] | 21782 | 64479 | 296 |
|    b. Sales[d] | 45308 | 202083 | 446 |
|    Subtotal (a + b) | 67090 | 266562 | 397 |
| Total inward (1 + 2 + 3) | 139938 | 468105 | 335 |

*Notes:*
[a]  Travel, passenger fares and other transport.
[b]  Petroleum industry and manufacturing services, real estate, wholesale and retail trade, hotels and public utilities excluding telecommunication.
[c]  Other private services; insurance is on a gross basis (premiums).
[d]  Excluding services delivered mainly via FDI listed in note (b).

*Source:    Survey of Current Business*, various issues.

*General note:*    Data on sales in this table are not fully comparable with those in Table 4.1, because they refer to sales to foreign persons for outward sales and sales to US persons for inward sales. They thus exclude sales back to the USA for outward transactions and sales to foreign persons for inward transactions. Excluded sales accounted in 1999 for 5.3 per cent of outward sales and 7.5 per cent of inward sales. It should also be noted that sales data are based on different industry classification in 1986–7 (SIC) and in 1999 (NAISC). The new industry classification resulted in the increase in sales of services at the cost of sales of goods.

*Table 4.3    Ratios of FDI sales to trade for selected services, United States, 1986–99*

| Service industry/category | Outward transactions ratio | | | Inward transactions ratio | | |
|---|---|---|---|---|---|---|
| | 1986 | 1990 | 1998 | 1987 | 1990 | 1999 |
| Finance | 2.7 | 2.4 | 2.8 | 0.9 | 2.2 | 4.5 |
| Insurance | 4.5 | 6.2 | 6.3 | 3.3 | 3.2 | 3.7 |
| Telecommunication | 0.1 | 0.1 | 2.5 | 0.1 | 0.1 | 2.9 |
| Construction, engineering and mining | 3.1 | 3.7 | 2.1 | 5.3 | 17.3 | 12.1 |
| Transport | 0.1 | 0.1 | 0.3 | 0.1 | 0.2 | 0.3 |
| Advertising | 20.5 | 31.0 | 13.8 | 10.3 | 11.2 | 6.0 |
| Computer and data processing | 3.9 | 7.5 | 19.4 | 9.9 | 46.8 | 3.7 |
| Accounting, R&D and management | 2.8 | 5.9 | 3.3 | 2.6 | 1.4 | 0.6 |
| Health | 1.0 | 0.6 | 0.3 | — | — | — |

*Source:    Survey of Current Business*, various issues.

with sales exceeding exports and imports by a factor of 14 and 6, respectively. Insurance is another industry with dominance of FDI relative to exports; this suggests that, although insurance technology is probably capable of being standardized and services delivered through information transfers, factors such as the need for trust to win over customers as well as import restrictions have so far made local establishment necessary or desirable for the provision of insurance services. The financial services industry shows visibly different trends with respect to internationalization of US firms and those of firms from other countries as reflected in their transactions in the USA: the ratio of US foreign affiliate sales to exports is, on the whole, stable or declining, while the ratio of foreign affiliates sales in the USA to US imports shows a noticeable increase. And finally, in telecommunication services, which were delivered in the second half of the 1980s predominantly via the trade mode, and the tradability of which further increased as a result of technological advances, the liberalization of FDI has shifted the modes decisively towards FDI.

**Intra-firm Trade as Evidence of Integrated Production**

As discussed earlier, there are a number of examples suggesting that international production of services is undertaken increasingly by both services- and goods-producing TNCs in an integrated manner, under simple or complex integration strategies, following a path taken much earlier by TNCs as regards international production of goods.

Can the scope of, and trends in, integrated international production of services be measured by the trade and foreign affiliate sales data? In general, intra-firm trade is an essential feature of all international production through FDI (UNCTAD, 1996, p.103). Its three principal streams – parents' exports to affiliates, affiliates' exports to parents and affiliate-to-affiliate exports – have been used as indicators of integrated production of goods and its various aspects. According to the *World Investment Report 1996*, 'the volume of intra-firm flows tends to increase, and the direction and nature of intra-firm flows and their geographic spread to change, with the complexity of corporate integration. Indeed, such flows are an indicator of the degree of integration of production . . . within TNCs' (UNCTAD, 1996, p.103). Simple integration strategies generate intra-firm trade between parents and affiliates. The location of a foreign affiliate upstream in the value chain leads to exports by the affiliate to the parent (or imports by the parent from the affiliate). Examples include foreign affiliates in natural resources or in labour-intensive manufacturing processes, located in developing countries. Foreign affiliates can also be allocated downstream functions, leading to exports from parents to their affiliates. Examples include assembly affiliates or trading affiliates, the former assembling and the latter selling abroad goods supplied by parent firms. Downstream integration of the trading function is widely used by non-US TNCs in selling goods to the US market.[20]

Complex strategies lead to greater multiplicity of linkages, resulting in greater intensity of affiliate-to-affiliate intra-TNC exports on a regional or even a global basis. In fact, inter-affiliate trade (exports of one foreign affiliate being imports of another one) signal the emergence of truly integrated international corporate systems. Thus intra-firm trade is a good indicator of integrated production, although in the case of affiliates' exports to parents it may point to either simple or complex integration, while in the case of affiliate-to-affiliate exports it more properly measures complex integration. Parents' exports to affiliates are the most ambiguous. They take place under traditional non-integrated strategies in the form of flows of intermediate goods and services from parent firms to stand-alone affiliates. But, as mentioned above, they can also be an indicator of simple integrated strategy. Furthermore, the intensity of parent-to-affiliate flows increases as TNCs move towards complex integrated strategies and the task of managing and coordinating TNC networks (comprising, increasingly, not only foreign affiliates but also firms associated with the network through alliances) becomes more complex.

As countries begin to publish data on intra-firm transactions in services, it becomes possible to examine the extent of, and trends in, various types of international production of services. In the USA, data on intra-firm trade in services are reported for two service categories: 'other private services', comprising business, professional, financial and telecommunication services, and 'fees and royalties' (not included in this discussion). As regards the former category, it appears that the importance of intra-firm trade in services is almost similar to that of intra-firm trade in goods, when both are compared with total trade in the respective categories at the country level. In 2000, the share of intra-firm exports in total US exports of other private services was 30 per cent and the share on the import side was 46 per cent. (Corresponding shares for intra-firm trade in goods for the latest year available, 1994, were 36 per cent for exports and 43 per cent for imports!)[21] On the export side, this share has not changed since 1986 (Figure 4.1). The shares of intra-firm exports accounted for by US parents'

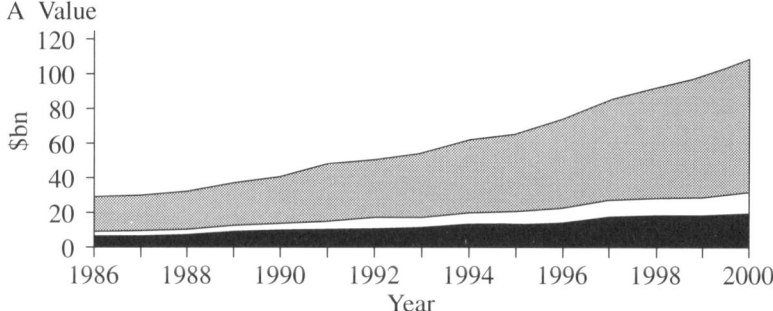

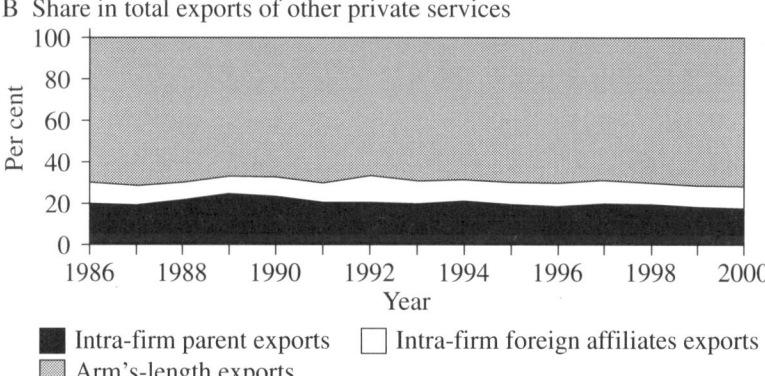

*Source:*   Authors' calculations, based on *Survey of Current Business*, various issues.

*Figure 4.1    Intra-firm exports of other private services, value and share in total exports, United States, 1986–2000 (billions of dollars and per cent)*

exports to their foreign affiliates and exports of foreign affiliates in the USA to their parents have remained stable during this period at the levels of, respectively, around 20 per cent and 10 per cent. On the import side there has been a considerable increase in intra-firm trade, which rose from 30 per cent of total imports in 1986 to 46 per cent in 2000 (Figure 4.2). Both imports by the US parent firms and those by foreign affiliates in the USA were responsible for this increase. In 2000, they each accounted for half of the US intra-firm imports of other private services.

It is difficult to draw any major inferences regarding intra-firm trade by individual service categories because of data limitations. From the data that are available, it appears that, at the end of the 1990s and in 2000, among

A  Value

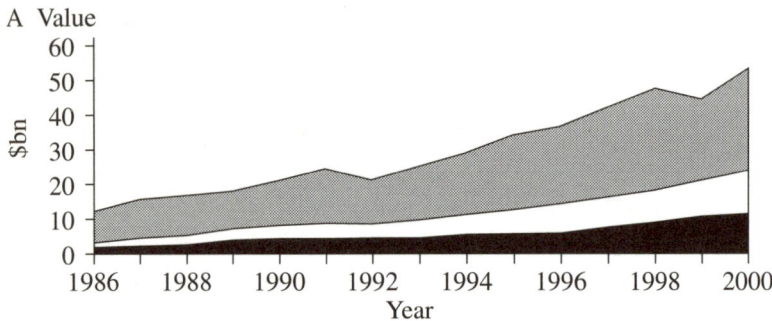

B  Share in total imports

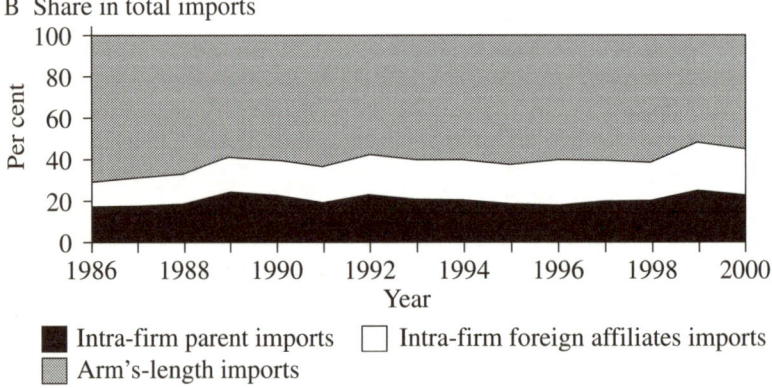

■ Intra-firm parent imports   □ Intra-firm foreign affiliates imports
▨ Arm's-length imports

*Source:*  Authors' calculations, based on *Survey of Current Business*, various issues.

*Figure 4.2    Intra-firm imports of other private services, value and share in total imports, United States, 1986–2000 (billions of dollars and per cent)*

the few (four to five) disaggregated categories of services, financial services were a fairly important area for trade in both directions between parent firms and their affiliates for both US and non-US TNCs.[22] Film and television tape rentals are an important item in exports of US parents to their affiliates and imports of foreign affiliates in the USA from their parents. Apparently, TNCs in the film industry use foreign affiliates as an international distribution channel for films and television programmes, an example of downstream simple integration into distribution similar to using marketing affiliates to sell goods. Computer and data-processing services are important as far as US parents' intra-firm exports are concerned, but not for their imports. But by far the largest category for all types of intra-firm trade is a broad category of 'other services' accounting for between 60 and over 85 per cent of the four trade streams between parents and affiliates. As regards parents' exports, these are services carried out by parent firms on behalf of all TNCs (management, coordination, R&D, and so on), the cost of which is typically divided among various units of TNCs, including foreign affiliates.[23] One can only speculate that, as regards service TNCs, they reflect more the support of parent firms for their still relatively stand-alone foreign affiliates than integrated strategies. But they also include intra-firm exports of services of goods-producing TNCs, much more advanced in pursuing integrated strategies than services TNCs. In this case, parents' exports of services to affiliates can be an indicator of integrated strategies. The lack of disaggregated data for this category makes it impossible to make further inferences about the nature of intra-firm trade in services.

Judging from data on imports of 'other private services' into the USA, intra-firm trade is more important in trade between developed countries than between developed and developing countries. However, the gap between the two in this respect narrowed somewhat over 1986–2000 (Table 4.4). In 2000, intra-firm imports from developed countries accounted for 57 per cent of total imports from these countries (37 per cent in 1986) while the share for imports from developing countries was 27 per cent (12 per cent in 1986). But for some developing countries these shares are much higher: for Hong Kong (China), over 80 per cent, for Singapore, 70 per cent; and for the Republic of Korea, China and Taiwan Province of China, over 40 per cent each.[24] The marked increase in the importance of intra-firm imports of services in total imports of services (from both developed and developing countries) into the USA suggests that in some service activities of particular importance to the USA, there may be a tendency for US TNCs to pursue integrated production strategies; that is, to buy services from specialized foreign affiliates located overseas.

The above comparisons focused on the role of international production

Table 4.4   *Imports of other private services into the United States, intra-firm and arm's-length, 1986 and 2000 (billions of dollars and per cent)*

| | 1986 | | | | | 2000 | | | | |
|---|---|---|---|---|---|---|---|---|---|---|
| | Total | Intra-firm total | Intra-firm US parents | Intra-firm US affiliates | Arm's-length | Total | Intra-firm total | Intra-firm US parents | Intra-firm US affiliates | Arm's-length |
| Developed countries, value | 8.75 | 3.24 | 1.96 | 1.34 | 5.53 | 35.39 | 20.14 | 9.87 | 10.27 | 15.26 |
| Composition | 100% | 37% | 22% | 15% | 63% | 100% | 57% | 28% | 29% | 43% |
| Developing countries,[a] value | 4.72 | 0.57 | 0.39 | 0.21 | 4.26 | 19.05 | 5.17 | 3.12 | 2.07 | 13.83 |
| Composition | 100% | 12% | 8% | 4% | 90% | 100% | 27% | 16% | 11% | 73% |
| All countries, value | 13.9 | 3.9 | 2.4 | 1.5 | 10.0 | 54.7 | 25.3 | 13.0 | 12.3 | 29.4 |
| Composition | 100% | 28% | 17% | 11% | 72% | 100% | 46% | 24% | 22% | 54% |

*Note:*   [a] Including countries in Central and Eastern Europe.

*Source:*   *Survey of Current Business*, various issues.

of services, including integrated production, in total transactions in comparable service categories at the country level. Such comparisons do not say much about whether, in the area of services production, corporate strategies evolve, as a result of increased tradability of services and opening of service industries to both trade and FDI, in the same manner as strategies in the area of (always tradable) goods evolved in the past, alongside the liberalization of trade and FDI in goods. As regards goods, between the late 1970s and early 1990s, the share of intra-firm exports and imports in total exports and imports did not change (Zeile, 1997, p.25) but, at the same time, integrated production of goods progressed greatly, as indicated by the increasing shares of affiliate-to-affiliate exports in both *total intra-firm trade* of US TNCs (comprising parents' exports to affiliates, affiliates' exports to parents and affiliate-to-affiliate exports) and in *total intra-firm exports of affiliates* (comprising the latter two categories), as shown in Table 4.5. Trade with other foreign affiliates was particularly intensive for US affiliates located in developed countries, and especially those in the European Union, where integrated strategies were vigorously pursued by US TNCs.

As regards affiliate-to-affiliate trade in services, it almost doubled in absolute value between 1986 and 1999 (Table 4.5),[25] indicating that integrated strategies in the production of services are indeed progressing, reflecting in the trade data the kind of transactions illustrated by examples of individual firms in the preceding section. Affiliate-to-parent exports, reflecting both simple and integrated strategies, also doubled. But both types of transactions increased at a slower pace than parents' exports to affiliates, which tripled during the same period. This suggests that US TNCs are not shifting, on balance, towards integrated international production strategies in services that involve locating service activities abroad at various points on the value chain in order to enhance efficiency of their entire systems. But the absolute volume of import transactions related to these strategies is growing fast enough to increase the importance of these transactions relative to total imports of other private services into the USA.

Another indicator of the importance of integrated strategies is the role of intra-firm transactions in total sales of foreign affiliates, comprising local sales in host countries, exports back to the home country and exports to third countries. Data on all three types of sales are available for arm's-length transactions and intra-firm transactions for MOFAs of US TNCs (Table 4.6). Between 1986 and 1999, all types of sales including intra-firm and arm's-length transactions increased considerably, between two times and more than five times. But, invariably, local sales grew faster than exports and arm's-length transactions grew much faster than intra-firm transactions. As a result, the share of local sales in total sales increased

*Table 4.5    The importance of affiliate-to-affiliate trade in intra-firm trade,*
*United States, 1977–99 (billions of dollars and per cent)*

|  | 1986 | | 1999 | |
|---|---|---|---|---|
| Category of intra-firm exports | Value $bn | % share | Value $bn | % share |
| A.  US TNCs' intra-firm exports of services |  |  |  |  |
| 1. Parents to affiliates | 5.6 | 28 | 18.3 | 40 |
| 2. Affiliates to parents | 7.9 | 39 | 14.5 | 32 |
| 3. Affiliates to affiliates | 6.8 | 33 | 12.6 | 28 |
| 4. US TNCs total intra-firm | 20.3 | 100 | 45.4 | 100 |
| B.  Affiliates' exports of services from host countries |  |  |  |  |
| 1. Total exports to US and 3rd countries | 21.9 | 100 | 48.6 | 100 |
| 2. Exports to other affiliates | 6.8 | 31 | 12.6 | 26 |

MEMORANDUM

| | 1977 | 1983 | 1993 |
|---|---|---|---|
| Share of affiliate-to-affiliate exports of goods[a] in total intra-firm exports of US TNCs, per cent | | | |
| | 30 | 40 | 44 |
| Share of affiliate-to-affiliate exports of goods[a] in total host country exports of US MOFAs, per cent | 1977 | 1983 | 1993 |
| | 37 | 53 | 60 |

*Note:*  [a] denotes that data refer to total sales but they are illustrative of the sales of goods which account for over 85 per cent of the total sales. For example, in 1994, the share of goods in total sales of MOFAs was nearly 90 per cent (US Department of Commerce, 1998, p.177).

*Source:*   Data on services exports: authors' calculations based on *Survey of Current Business*, various issues; data on total exports of goods and services (UNCTAD, (1996, p.105).

from 73 per cent in 1986 to 86 per cent in 1999, while the share of sales to unaffiliated persons increased during the same period from 75 to 90 per cent.[26] This suggests that the influence of factors requiring proximity to clients and presence in host country markets is greater than that of factors pushing towards international production of services in an integrated manner.

*Table 4.6  Transactions of foreign affiliates in host countries, growth and composition (billions of dollars and per cent)*

| | Value (billions of dollars) | | Increase (per cent) 1999/1986 | Composition (per cent) | |
|---|---|---|---|---|---|
| | 1986 | 1999 | | 1986 | 1999 |
| A  Sales of MOFAs of US TNCs | | | | | |
| 1. Total sales (1 = 2 + 3) | 82.6 | 357.3 | 433 | 100 | 100 |
| 1a. to affiliated persons (1a = 2a + 3a) | 19.6 | 36.9 | 188 | 24 | 10 |
| 1b. to unaffiliated persons (1b = 2b + 3b) | 63.0 | 320.4 | 509 | 76 | 90 |
| 2. To US persons (exports from host countries to the US) | 9.8 | 18.9 | 193 | 12 | 5 |
| 2a. to US parents | 7.9 | 14.5 | 184 | 10 | 4 |
| 2b. to unaffiliated US persons | 1.9 | 4.4 | 232 | 2 | 1 |
| 3. To foreign persons (3 = 4 + 5) | 72.8 | 338.4 | 465 | 88 | 95 |
| 3a. to other foreign affiliates (3a = 4a + 5a) | 11.7 | 22.4 | 191 | 14 | 6 |
| 3b. to unaffiliated foreigners | 61.2 | 315.9 | 516 | 74 | 88 |
| 4. Local sales | 60.7 | 307.7 | 507 | 73 | 86 |
| 4a. to other foreign affiliates | 4.9 | 9.8 | 200 | 6 | 3 |
| 4b. to unaffiliated foreigners | 55.8 | 297.9 | 534 | 68 | 83 |
| 5. Sales to other countries (exports to third countries) | 12.1 | 30.7 | 254 | 15 | 9 |
| 5a. to other foreign affiliates | 6.8 | 12.6 | 185 | 8 | 4 |
| 5b. to unaffiliated foreigners | 5.3 | 18.1 | 342 | 6 | 5 |
| B  Sales by foreign affiliates in the USA [a] | | | | | |
| 1. Total sales | 66.3 | 312.9 | 472 | 100 | 100 |
| 2. Local sales (to US persons) | 62.6 | 289.3 | 462 | 94 | 92 |
| 3. Exports (sales to foreign persons) | 3.8 | 23.6 | 621 | 6 | 8 |
| 3a. to foreign parents | 1.6 | 10.5 | 656 | 2 | 3 |
| 3b. to foreign affiliates | 0.2 | 1.2 | 600 | 0 | 0 |
| 3c. to other foreigners | 1.9 | 12.0 | 632 | 3 | 4 |

*Note:*  [a] Data in the first and fourth columns are for 1987. The increase in the third column refers to 1999/1987.

*Source:  Survey of Current Business, various issues.*

## CONCLUSION

Technological developments in services are working towards increased internationalization of the services sector through FDI as well as trade. Improved firm-specific advantages due to advances in service technologies have enhanced the abilities of firms to compete with local firms in foreign markets, whether through FDI, or trade, or a combination of the two. At the same time, telecommunication and data technologies have increased the tradability of some services, creating new opportunities for trade, especially in information-intensive services. Furthermore, liberalization of FDI regimes, driven partly by the need for access to sophisticated and improved service products in order to maintain growth and competitiveness, has expanded the scope for international delivery of services through the establishment of local operations in services that require close interaction between provider and customer or physical proximity between the two.

Together, these developments have created scope not only for increased trade and FDI in services to serve customers, but also for the pursuit of integrated international production strategies by TNCs. Depending upon the nature of the service products at various points along their value chain of a final good or service, firms can combine FDI and trade to take advantage of cost differences between locations and scale economies to distribute different activities or functions in the most efficient manner. These factors create increasing complementarity between trade and FDI. It is difficult, however, to predict a priori the direction of the resulting changes in the modes of internationalization of services at the sectoral or even industrial level. The effect of technological change on the relative importance of FDI and trade and on intra-firm trade generated by integrated international production for services taken as a whole or broad service categories will depend on how technological change affects different services as regards the tradability of products and the firm-specific, internalization and locational advantages determining FDI.

Judging from comparisons based on US data, the internationalization of services as a whole and of a majority of individual service industries now takes place more through FDI than through trade, and the relative importance of FDI as a mode of international delivery of services has increased in recent years. The data also suggest that the overwhelming proportion of total services sales through FDI (foreign affiliates' sales) are for local (host country) markets rather than for exports from host countries. The proportion of local sales in total sales by foreign affiliates rose noticeably between the late 1980s and the late 1990s as far as US TNCs operating abroad are concerned. It has always been very high for the sales of foreign affiliates of other countries' TNCs in the USA. These trends suggest that, overall, the impact of technological developments affecting services production and delivery has been more to strengthen the ability of firms to compete internationally through the establishment of operations in foreign markets than to strengthen their ability to compete through trade, defined broadly as transactions between residents and non-residents (and thus not making the distinction between the three GATS non-FDI modes of delivery). However, within the category of services that can be delivered by both modes of delivery, trade has increased rapidly, along with FDI-related sales, pointing to increased complementarity of trade and FDI rather than FDI replacing trade.

TNCs in services and goods use FDI in services for many strategies, including integrated production strategies, marketing strategies and management strategies. It seems, from the data available, that the former, although growing in significance, are not the most important. In relative terms, their importance may even be declining, as evidenced by the declining shares of intra-firm transactions in total transactions of foreign affiliates in host countries, and especially of affiliate-to-affiliate trade in total intra-firm trade of TNCs. However, judging from the value of intra-firm imports of services relative to total imports of services at the country level, the importance of trade generated by integrated strategies in total trade appears to be increasing. This suggests that TNCs in the USA and, perhaps, other developed countries are using their international production networks to exploit cost differences in services production across countries more fully to improve efficiency and competitiveness, but the scope of integrated production is as yet limited. Moreover, as the trend towards externalizing and outsourcing of services to other firms by firms within countries remains strong, it is possible that it will become more important internationally and that the growing importance of FDI in services will be supported and complemented increasingly by transactions between firms through contractual arrangements and alliances of various kinds, rather than by transactions between members of TNCs' own production systems.

## NOTES

1. Views and opinions expressed in this chapter are entirely the authors' own and cannot be attributed to the organization with which they are associated.
2. See UNCTAD (1996, ch. IV).
3. See UNCTAD (1993, chs V and VI), and UNCTAD (1996, ch. IV).
4. The production network for Ford's Fiesta, for instance, involved affiliates in 12 locations spread over four countries (France, Germany, Spain and the United Kingdom) producing different components, with three of them involved in final assembly (Dicken, 1992, p.300).
5. Diesel engines were manufactured in and exported from Thailand, transmissions from

the Philippines, steering gears from Malaysia and engines from Indonesia (UNCTAD, 1996, p.100).

6. Information on American Express, General Electric, Hong Kong and Shanghai Bank, and Telefonica in the examples below is from 'Globalization Goes Upscale', *The Wall Street Journal Europe*, 4 February 2002.

7. Information obtained from NFIA-Company Profiles: 'Cisco Systems' (*http://www.nfia,com/html/company/c_cs/html*, 3/29/2002); 'SingAREN and Cisco collaborate to deliver next generation broadband network' (jointly issued by SingAREN and Cisco Systems, Inc.) (*http://www.krdl.org.sg/GeneralPressClips/singarenl.html*. 3/29/2002); and 'Cisco CEO to visit India in January', *ZDNetIndia News* (*http://www.zdnetindia.com/news/breaknews/stories*).

8. 'Company town keeps Indians at home', *The New York Times*, 18 March 2002.

9. 'Intel chief opens R&D labs with Infosys', *The Industry Standard* (*www.thestandard.com/wire*, 3/29/2002).

10. For example, the functional and geographic scope of Cisco Systems' international activities is wide enough to suggest a potential for the pursuit of integrated international production strategies comparable to those of TNCs in some manufacturing industries.

11. *The Economist*, 1 December 2001, p.59.

12. See Auguste *et al.* (2000).

13. For example, Infosys (India) provides software coding and maintenance and software applications to clients abroad ('Globalization Goes Upscale', *The Wall Street Journal Europe*, 4 February 2002, and 'Intel chief opens R&D lab with Infosys', *The Industry Standard*, 3/29/2002 at *http:// www.thestandard.com/wire*). The Ghana office of ACS, a US-based firm, provides data processing for Aetna, the US health insurance firm ('In Ghana, Hope Arrives via Satellite', *International Herald Tribune*, 9 May 2001).

14. EDS's activities include systems development, integration and management and a range of other innovative offerings.

15. One typology of services divides services by decreasing need for customer physical proximity into three categories: those involving tangible actions to customers in person (for example, passenger transport, healthcare, airlines), for which local presence is important; those involving tangible actions to physical objects to improve their value to customers (for example, freight transport, warehousing, equipment installation and maintenance), for which local presence is required but service processes can often be administered from a distance; and information-based services, dependent on collecting, manipulating, interpreting and transmitting data to create value (for example, banking, insurance, consulting), potentially able to be delivered from anywhere, with local presence limited to a terminal. It is noted that, within each category, there is a tendency for the back-office elements of the service to be less proximate to the customer group, while the front office is still subject to pressure for customer proximity (see Hood and Peters, 2000, pp.88–9).

16. With respect to the differential patterns of technological innovation, service categories that have been identified as technology-intensive include scale-intensive physical and information network-based services, science-based and specialized supplier services which are actively engaged in the development and use of data, communication and storage and transmission of information (Miozzo and Soete, 2001, p. 163). Examples include banking and insurance, stock exchange institutions, cellular phone services, office automation, engineering design, express package transportation and airline reservations.

17. Thus trade includes payments and receipts for travel, passenger fares, other transportation and other private services. It excludes fees and royalties because they refer to specific modes of delivery of certain services and goods (so-called 'non-equity modes' such as licensing and franchising) rather than to trade in these services. Fees and royalties include only a fraction of the total value of sales of services produced under these modes.

18. In 1980, US TNCs accounted for half of the world stock of FDI in services (Mallampally and Zimny, 2000, pp.50–1).

19. Another reason for underestimation of sales is that 'the estimates of cross-border

exports and imports include services provided by banks, whereas those of sales through affiliates cover non-bank affiliates only' (US Department of Commerce, 2001, p.51).

20. As a result, wholesale trading affiliates accounted for two-thirds of the intra-firm imports of foreign affiliates in the USA in 1994. Non-US TNCs also use widely trading affiliates in the USA to import products from the USA. These affiliates accounted for two-thirds of foreign affiliates' intra-firm exports from the USA. This is an example of upstream integration of the trading (or purchasing) function (Zeile, 1997, pp.28–9).
21. Zeile (1997, p.25).
22. US Department of Commerce (2001, pp.62–3).
23. Ibid., p.63.
24. Ibid., p.83.
25. Affiliate-to-affiliate transactions within the same host country (local sales), shown in Table 4.6, also doubled, from $5 billion to $10 billion. They also reflect division of labour between affiliates and, thus, integrated production, but not giving rise to international trade.
26. In the case of other countries' transactions in services, judging from data on foreign affiliates' sales in the USA, the relative importance of local sales in total sales of foreign affiliates has not risen, but the ratio remains extremely high (94 per cent in 1987 and 92 per cent in 1999).

# REFERENCES

Auguste, B.G., Y. Hao, M. Singer and M. Wiegand (2000), 'The other side of out-sourcing', *The McKinsey Quarterly*, 1 (*http://www.mckinseyquarterly.com*).

Dicken, P. (1992), *Global Shift: The Internationalization of Economic Activity*, New York: The Guilford Press.

Dunning, J.H. (1989), *Transnational Corporations and Growth of Services: Some Conceptual and Theoretical Issues*, UNCTC Current Studies, series A, no. 9, New York: United Nations.

Hood, N. and E. Peters (2000), 'Globalization, corporate strategies and business services', in N. Hood and S. Young (eds), *The Globalization of Multinational Enterprise Activity and Economic Development*, London and New York: Macmillan Press and St Martin's Press.

Kravis, B.I. and R.E. Lipsey (1988), 'Production and trade in services by US multinational firms', NBER Working Paper no. 2615, New York.

Mallampally, P. and Z. Zimny (2000), 'Foreign direct investment in services. Trends and patterns', in Y. Aharoni and L. Nachum (eds), *Globalization of Services. Some Implications for Theory and Practice*, London and New York: Routledge.

Miozzo, M. and L. Soete (2001), 'Internationalization of services: a technological perspective', *Technological Forecasting and Social Change*, **67** (2), 159–85.

Sauvant, K.P. (1990), 'The tradability of services', in P.A. Messerlin and K.P. Sauvant (eds), *The Uruguay Round. Services in the World Economy*, Washington, DC and New York: The World Bank and United Nations Centre on Transnational Corporations (UN CTC).

Sauvant, K.P. and Z. Zimny (1987), 'Foreign direct investment in services: the neglected dimension in international service negotiations', *World Competition*, **31**, 27–55.

UNCTAD (1993), *World Investment Report 1993: Transnational Corporations and Integrated International Production*, New York and Geneva: United Nations.

UNCTAD (1994), *The Tradability of Banking Services: Impact and Implications*, Geneva: United Nations.

UNCTAD (1996), *World Investment Report 1996: Investment, Trade and International Policy Arrangements*, New York and Geneva: United Nations.

UNCTAD (1999), *World Investment Report 1999: Foreign Direct Investment and the Challenge of Development*, New York and Geneva: United Nations.

UNCTAD (2001), *World Investment Report 2001: Promoting Linkages*, New York and Geneva: United Nations.

UNCTC (1989a), *Transnational Service Corporations and Developing Countries: Impact and Policy Issues*, New York: United Nations.

UNCTC (1989b), *Foreign Direct Investment and Transnational Corporations in Services*, New York: United Nations.

US Department of Commerce (1998), *US Direct Investment Abroad: 1994 Benchmark Survey, Final Results*, Washington, DC: US Department of Commerce.

US Department of Commerce (2001), 'US International Services: Cross-Border Trade in 2000 and Sales Through Affiliates in 1999', *Survey of Current Business*, November 2001, pp.49–95.

Zeile, W.J. (1997), 'US Intra-firm Trade in Goods', *Survey of Current Business*, February 1997, pp.23–38.

PART III


International Service Multinationals and the
Location of Production and Innovation
Activity

# 5. Globalization, regionalization and 'scales of integration': US IT industry investment in Southeast Asia

## Neil M. Coe

## INTRODUCTION

This chapter explores the spatial complexity of transnational corporation (TNC) organizational structures through an investigation into the growth strategies of US-based IT firms in the Asia–Pacific and, in particular, Southeast Asia.[1] It emerges from the continued unease some have with current definitions and conceptualizations of processes of international corporate expansion and, in particular, their applicability to processes of service activity internationalization. In definitional terms, there is still not enough clarity as to what terms such as regionalization, internationalization and globalization actually mean. Dicken *et al.* (1997), for example, argue that precise definitions are crucial to avoiding oversimplistic caricatures of complex processes. Equally, 'fuzzy' interpretations of these terms are a barrier to productive and comparable empirical research, and any subsequent policy or strategy recommendations (Markusen, 1999). In conceptual terms this chapter will make the argument that the investment strategies of TNCs are far more spatially variable and complex than is suggested in much of the contemporary international business literature. As Dicken *et al.* (1997, p.163) argue, 'the real point is to recognize the diversity and complexity of the processes and structures involved as firms increasingly operate across national boundaries'.

In addition, a further aim of the chapter is to consider how such conceptualizations might apply to service activities, which often do not conform to the typical manufacturing model – namely the sequential adding of value in different locations as products are assembled – that seems to underlie much of the literature on this topic. It has become an often repeated, but nonetheless true, observation that service internationalization is poorly understood in relation to that of primary and manufacturing activity. As

Greenwood *et al.* (1999) describe, services – and in particular business and producer services – differ in two key ways. Firstly, they deal primarily in knowledge, and hence overseas growth is relatively more dependent on the mobility of employees. Secondly, production and marketing are more closely tied, despite the impact of forms of IT that are making some services more 'tradable' (Daniels, 1993), resulting in more extensive patterns of corporate activity. There seems to be an assumption in some literature, however, that this leads to a replicated multidomestic structure of national subsidiaries. This chapter aims to show that in reality the picture is far more complex than this, with service activities being coordinated and organized on intermediate 'regional' scales between the 'national' and the 'global'.

The general context for this chapter is a world economy that is seeing progressively more foreign direct investment (FDI) poured into Southeast Asia and other parts of the Asia–Pacific. In 2000, South, East and Southeast Asia accounted for 18·8 per cent of the global stock of FDI, up from 15·7 per cent in 1990. In the year 2000 the region received US$137.3bn worth of inward investment, compared to an average of US$35.1bn a year over the period 1989–94 (UNCTAD, 2001). A notable proportion of this inward investment is accounted for by the activities of US-owned TNCs. In 1999, the USA generated US$118·6bn of outward FDI, of which US$30.5bn went to the Asia–Pacific region: Singapore alone accounted for US$6.3bn (Department of Commerce, July 2000). By that year, the United States had a cumulative investment of US$24·8bn in Singapore, of which US$9.9bn was in industrial machinery and electronics, and US$8·1bn in finance and real estate. A very small but rapidly increasing component of this investment is accounted for by the computer services and software sector.

The chapter will be structured in the following way. The next section presents a framework for analysing the international growth of TNCs that explicitly recognizes the increasingly 'regional' structure of corporate formations before, secondly, some brief comments on methodology are offered. The third substantive section of the chapter provides an overview of US IT industry investment into Southeast Asia. This provides empirical weight to the suggestion that, especially for clients facing 'service' functions, TNCs are increasingly instituting regional structures to mediate between their global and local/national spheres of activity. The fourth section seeks to add a level of complexity to the argument by building on this notion of regional structures to consider three key dimensions of spatial variation in corporate organizational forms. In essence, corporate structures differ greatly between and within both the various regional formations, and different kinds of economic activity. While this seems a commonsense observation, these are important variations that are overlooked in many aspatial and simplifying models of international corporate expansion. The conclusion considers some of the key implications of this analysis, both for understanding corporate globalization, and with respect to the other key focus of this volume, processes of innovation, of service activities.

## CONCEPTUALIZING PATTERNS OF INTERNATIONAL BUSINESS

The key to conceptualizing effectively the patterns of international business is the recognition that the modern global economy is characterized by the increasingly complex functional integration of economic activities across national boundaries (Dicken, 1998). Complex integration strategies can be distinguished from simple forms of integration such as international subcontracting or outsourcing (UNCTAD, 1993), in that foreign affiliates may perform, either by themselves or with other affiliates, the parent firm or partner firms, functions for the firm as a whole. Such a strategy is built upon a willingness to locate various corporate activities in places where they can best fulfil the firm's overall strategy. The literature is replete with examples – often drawn from the automotive and electronics sectors (see Dicken, 1998, for a review) – of how manufacturing activity is increasingly integrated across national territories, and there is growing evidence that TNCs are locating a small but rising proportion of their research and development activity outside their home countries (see, for example, Howells and Wood, 1993). Far less is known, however, about the extent to which a range of other service functions are internationally functionally integrated. UNCTAD (1993) lists procurement, accounting, finance, training, corporate planning and legal services under this heading, but further service activities such as sales and marketing, partner and channel management, distribution and logistics, customer support, internal IT provision and support, and personnel services could all be usefully added to the list of activities that may be organized through integrated transnational networks. For firms the main activity of which is actually a service, the delivery and implementation of these services, although generally assumed to be organized locally through national subsidiaries, may actually be performed through teams of specialists organized on a regional or, occasionally, global scale (for example, in management consultancy).

There are a number of broad frameworks in the literature that intersect notions of the geographic expansion and functional integration of international business activity (for example, Ruigrok and van Tulder, 1995; Bélis-Bourgouignan *et al.*, 2000). The difficulty with this approach is that it is

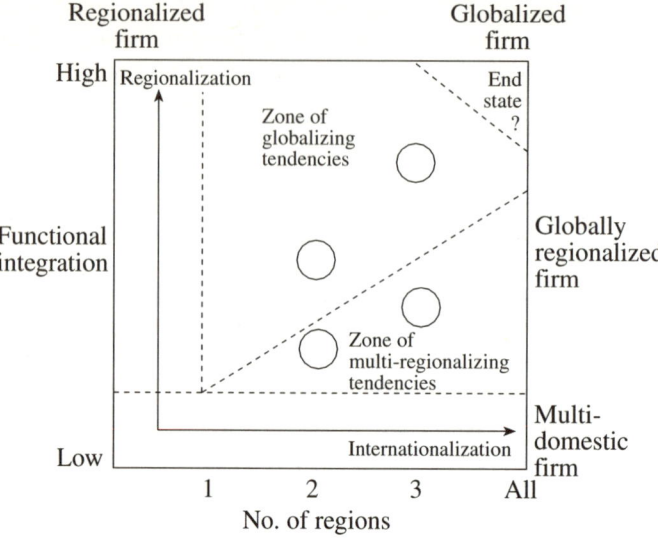

*Figure 5.1   Conceptualizing international expansion strategies*

hard to convey effectively the regional nature of many transnational organizational structures. Figure 5.1 represents an attempt to make sense of this
analytical space. This diagram builds upon the assumption that many
major firms are moving towards becoming 'globally regionalized'; that is,
they have constructed (or are constructing) integrated production and
delivery systems within each of the major triadic regions as they seek to
derive 'economies of regionalization' (Bartlett and Ghoshal, 1989; Dicken
*et al.,* 1997). As Yeung *et al.* (2001, p.158) describe, 'regions are becoming
the primary organizational focus of the world-wide web of TNC activities',
and this regionalization can be conceptualized 'as a strategic response to
the structural context of triadization and global competition and the need
for a regional structure to manage operations in different regions' (ibid.,
p.179). In other words, firms need to be able to adapt to conditions in what
may be very different host economies, and yet at the same time manage
effectively their global operations. For example, Rugman (2000, p.12) suggests that 'a paradox of globalization is that it never really occurred; it is a
myth. Instead, the vast majority of TNC manufacturing and service activity is (and always has been) organized regionally and not globally. While
TNCs are the engines of international business, their strategies are
regional' (see also Poon, 1997; Mirza, 1998).

Figure 5.1 maps out some of the key terms and processes. Simple regionalization represents a process of increasing functional integration within a contiguous regional space (for example, Western Europe, Southeast Asia). Hence regionalization is taken here to refer to corporate organization, rather than to political groupings (Lévy, 1995) or simply patterns of trade and FDI flows (Dicken and Yeung, 1999; Poon *et al.*, 2000). Internationalization, as Dicken (1998) has suggested, represents a simple international expansion of activity across national boundaries, perhaps resulting in some form of replicating multidomestic structure. Globalization, however, represents a simultaneous increase in both regional presence and functional integration. From this viewpoint, the 'globalized firm' is a notional end state in which a firm is present in all regions and functionally integrates its activities across all the territories. It is possible to identify an intermediate formation that is empirically very significant in patterns of international business, and which will be termed 'multi-regionalization'. Here, firms are expanding across regional territories and are integrating their operations within those regions, but to a much lesser extent across them (obviously, by definition, there will always be links of some kind back to the home region and ultimate headquarters). The notional end state here is a 'globally regionalized' firm that has relatively self-contained and integrated operations in all the major economic regions. At the moment the three key regions in such a structure are North America, Western Europe and East Asia or the Asia–Pacific, but Figure 5.1 allows scope for more regional structures to be developed. There are, of course, profound developmental implications for areas of the globe not integrated into these triadic structures (Stallings, 1995).

While this diagram is obviously grossly simplifying, it offers a useful starting framework in which to place the prevailing (or profile the changing) corporate strategy of a sector, firm or part of a firm. This last point is significant. It is often not possible to characterize the growth processes of an entire firm as simply conforming to internationalization or globalization: different kinds of activities within the corporation may be integrated or coordinated on different geographic scales. As Pritchard (2000, p.249) observes, 'at any time there are multiple (and perhaps contradictory) geographies at work within individual corporations'. Interpretations that consider TNCs as monolithic wholes, or that deal in ideal types such as the 'global corporation' (for example, Reich, 1991), simply do not do justice to the complexity of TNC organizational structures. In this chapter the term 'scales of integration' is coined to capture the way in which different corporate activities are controlled and coordinated at different spatial levels. By identifying and interpreting these multiple scales of organization within TNCs it is possible to avoid the attraction of unproblematically privileging the global scale in analyses (Fagan, 1997). While there is now a burgeoning literature in geography on the social and political construction of scale (see, for example, Smith, 1993; Swyngedouw, 1997; Cox, 1998), there has been far less work on how corporations 'construct' geographic scales through their organizational structures, and how these scales both reflect and reinforce functional spatial units within the global economy.

This is also an extremely important exercise in the sense that there are developmental implications for cities and regions that intersect with these organizational geographies in different ways. For example, there may be beneficial outcomes for places that attract the coordination and control functions, often in the form of regional headquarters (RHQs), of these complex and overlapping scales of integration. As TNCs increasingly organize their operations on the regional scale, RHQs may become necessary mid-level structures to implement global strategies at the regional level, and are an extremely important site in 'matrix' management structures, becoming responsible not simply for a product or area structure but for a 'whole set of dual reporting links between product and area segments of the TNC' (Yeung *et al.*, 2001, p.166). Regional offices tend to be established when the branch network becomes extensive in a region, the organization seeks regional or global functional integration, or when the firm management favours decentralized decision making. Such offices are seen as a highly desirable form of foreign direct investment, owing to the relatively high-quality direct employment they provide, the high levels of interaction with financial and business services, and their influence over investment decisions (Ho, 1998).

## METHODOLOGY

This analysis draws on 40 in-depth interviews undertaken with managers of US IT firms with bases in Singapore during the second half of 1999. It is important to note that the IT industry was defined broadly to cover firms that provide any form of IT product or service. In this sense no nominal IT manufacturing/service distinction was applied; rather, the research sought to uncover which different forms of activity were being undertaken in the region, and to characterize their relative importance and different geographies. Pre-assigning firms to different sectors in an arbitrary manner is increasingly seen to hamper analysis (Allen, 1992) as all economic activities combine both goods and service functions to different degrees. Firms selling personal computers, for instance, will usually provide customers with some form of maintenance and support service, and will need to use a range of corporate services themselves (for example, sales, accounting, distribution) to support their business. Overall, this in an industry in which there has been a marked shift over the last 20 years, from hardware to software and services revenues as margins on hardware products have fallen steadily and providers have sought to add value through a wide range of service and consultancy activities (Coe, 1997). The IT industry should thus best be seen as a continuum, running from hardware manufacturing at one end (goods-intensive), through pre-packaged software and tailored software, to IT consultancy (knowledge/service-intensive) at the other. Many firms undertake a range of these activities, making it hard to identify 'service' or 'manufacturing' companies: hence the use of the term service 'activities' rather than 'sectors' or 'firms' in this chapter. For the purposes of this study, the IT industry was not taken to include firms in electronics production chains, which have already been much studied in East and Southeast Asia (Wong, 1998; Mathews, 1999): the firms selected were those that deliver IT products and services directly to firms and consumers. The aim here was to bring a less studied and more recently established dimension of US foreign direct investment into view.

Singapore was chosen to undertake this research as it is home to the vast majority of US IT firms with bases in the region, being both a key market and organizational node within Southeast Asia and, in some instances, the Asia–Pacific. Directory searches revealed some 120 US IT firms in Singapore in early 1999, although this is almost certainly an underestimate due to the usual problems of attaining a complete, current list in such a rapidly developing sector. The sample was not stratified by size, rather it was designed to cover as many of the most significant firms as possible. Where firms could not be interviewed, basic data were obtained from a postal questionnaire and various secondary sources. Overall, the data provide a snapshot of the leading US IT industry investments in Singapore at a particular point in time. The key topic covered in the interviews was the organizational structure of the firms in Southeast Asia, and the relationships between these investments and broader structures at both the Asia–Pacific and global levels.

## PROFILING US IT INDUSTRY INVESTMENT IN SOUTHEAST ASIA

Table 5.1 provides an introductory picture of the leading US IT industry investments in Singapore. Several observations can be made at the aggregate level. Firstly, this is an extremely recent wave of inward investment, with many firms entering Singapore during the 1990s, although some broad-based vendors such as Hewlett Packard and IBM have had a presence in the city since the 1970s because of their hardware interests. Many of the surveyed firms are constructing a network of Asia–Pacific or

Table 5.1  US IT industry FDI into Singapore

| Firm | Singapore employees, mid-1999 | Core business | Regional HQ? | Year estab. Singapore | US HQ |
|---|---|---|---|---|---|
| 1. Hewlett Packard | 8500 | Hardware manufacturing, sales and distribution, R&D | SE Asia, Asia–Pacific | 1970 | Palo Alto, CA |
| 2. IBM | 4000 | Hardware manufacturing, sales and distribution, R&D | ASEAN | 1975 | Armonk, NY |
| 3. Compaq | 1465 | Hardware manufacturing, sales and support | ASEAN, Asia–Pacific | 1987 | Houston, TX |
| 4. CSC | 700 | Computer services | Asia | 1998 | El Segundo, CA |
| 5. 3Com | 500 | Networking products, manufacturing, sales and support | Asia–Pacific | 1990 | Santa Clara, CA |
| 6. EDS | 350 | Computer services | Asia | 1985 | Plano, TX |
| 7. Sun Microsystems | 300 | Hardware sales and support | South Asia | 1988 | Palo Alto, CA |
| 8. Oracle | 300 | Database software sales and support | Asia–Pacific, South Asia | 1988 | Redwood Shores, CA |
| 9. Andersen Consulting (now Accenture) | 250 | IT Consultancy | No | 1975 | New York |
| 10. Microsoft | 200 | Software manufacturing, distribution, sales and support | Asia–Pacific | 1990 | Redmond, WA |
| 11. JD Edwards | 150 | ERP software sales and support | Asia–Pacific | 1989 | Denver, CO |
| 12. Mastech | 150 | Computer services | South Asia | 1995 | Pittsburgh, PA |
| 13. Arthur Andersen | 140 | IT Consultancy (only Technology Division considered) | No | 1994 | Chicago, IL |
| 14. Cisco | 140 | Networking products, sales and support | Asia | 1995 | San Jose, CA |

124

| 15. Dell | 130 | Hardware sales and support | Asia–Pacific | 1995 | Austin, TX |
| 16. Sybase | 100 | Database software development | No | 1998 | Emeryville, CA |
| 17. Gateway | 80 | Hardware sales and support | No | 1997 | North Sioux City, SD |
| 18. Peoplesoft | 75 | ERP software sales and support | Asia | 1996 | Pleasanton, CA |
| 19. BMC Software | 66 | Software sales and support | Asia–Pacific | 1997 | Houston, TX |
| 20. Parametric technology | 55 | Software sales and support | No | 1992 | Waltham, MA |
| 21. Novell | 50 | Networking software sales and support | ASEAN | 1989 | Provo, UT |
| 22. Silicon Graphics | 40 | Hardware sales and support | South Asia | 1988 | Mountain View, CA |
| 23. Informix | 40 | Database software sales and support | Asia–Pacific | 1989 | Menlo Park, CA |
| 24. ACI Worldwide | 40 | Financial software sales and support | Asia–Pacific | 1990 | Omaha, NB |
| 25. Symix | 35 | ERP software sales and support | Asia–Pacific | 1994 | Columbus, OH |
| 26. Imation | 32 | Computer peripherals sales | South Asia | 1996 | Oakdale, MN |
| 27. Autodesk Asia | 31 | CAD software sales and support | No | 1990 | San Rafael, CA |
| 28. Hitachi Data Systems | 30 | Software sales and support | No | 1979 | Santa Clara, CA |
| 29. Cincom | 30 | ERP software sales and support | ASEAN | 1982 | Cincinnati, OH |
| 30. EXE Technologies | 30 | Software sales and support | Asia–Pacific | 1996 | Dallas, TX |
| 31. Aspen | 26 | Software sales and support | Asia–Pacific | 1994 | Cambridge, MA |
| 32. Cadence | 25 | Software sales and support | Asia–Pacific | 1994 | San Jose, CA |

*Source:* Author's research.

Southeast Asian operations extremely quickly, and the IT sector can be used as a powerful example of the potential speed of overseas expansion. Secondly, the majority of firms are undertaking service activities (in particular sales, marketing and support) in Singapore. By contrast, relatively little R&D is undertaken in Singapore or indeed the broader region – none in the majority of firms – thus re-emphasizing the now well-documented home region embeddedness of this core function. Equally, few firms are undertaking manufacturing production in Singapore or the region, with the exception of firms producing shrink-wrapped, non-customized software, such as Microsoft, and hardware firms such as Compaq, IBM and HP, the largest firms in employment terms. Much of this FDI is thus clearly market-oriented (Dunning, 1993), that is, intended to enable the effective distribution of products and services into Southeast Asia and the wider Asia–Pacific region. Many firms were obtaining between 12 and 15 per cent of their total revenues from their Asia-Pacific division.

Thirdly, the majority of firms exhibited some form of multiregional structure built around the Americas, Europe and the Asia–Pacific, with the Asia–Pacific region usually being the last to be developed (Ho, 1998). Figure 5.2 illustrates two typical organizational structures for US IT firms, one 'emerging', or indicative of firms with fairly low levels of sales into the region, and one more 'established', with two regional layers of coordination to manage an extensive range of offices. Of course, some firms may only ever achieve a limited presence in the region, but many develop the extent of their operations over time. This expansion into the Asia–Pacific can be related back to Figure 5.1. In essence, what is seen here is a process of multi-regionalization (from two to three regions) as the US IT firms add an Asian dimension to existing operations in North America and Western Europe.

Fourth, within this multiregional structure, a significant number of firms have established a regional headquarters (RHQ) function in Singapore. As Table 5.1 indicates, Singapore acts as a centre of control and coordination for many US IT firms in Southeast Asia, in some instances for the whole Asia–Pacific, and in three cases – Hewlett Packard, Compaq and Oracle – both. These regional offices in the IT industry are part of a much wider, and increasingly well-studied, trend (for example, Daniels, 1987; Dicken and Kirkpatrick, 1991; Forsgren *et al.,* 1995). Several centres of corporate control have emerged in the Asia–Pacific region to support ever-increasing levels of inward foreign direct investment. Singapore is one of the two leading centres for RHQs in the Asia–Pacific, with Hong Kong its main competitor, although other centres such as Kuala Lumpur, Bangkok, Jakarta, Seoul and Taipei are also starting to emerge (*The Economist,* 3 April 1999). A survey by Perry *et al.* (see Perry, Poon and Yeung, 1998;

(a) Emerging structure

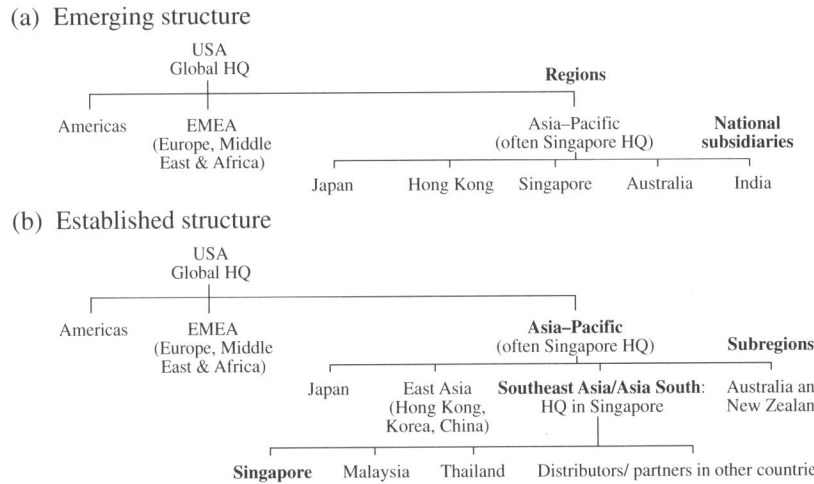

(b) Established structure

*Figure 5.2   Typical organizational structures of US IT industry TNCs*

Perry, Yeung and Poon, 1998) estimated that there was a population of some 600 RHQs in Singapore in the mid-1990s, defined as offices with responsibilities for the operation of one or more affiliates or subsidiaries outside Singapore.

The key functions of these offices in Singapore tend to be control and coordination, sales and service support, financial management, strategic planning and training. The island's attractiveness to foreign business is well chronicled, with Perry *et al.*'s survey identifying the following five most important factors: its centrality to the region, the general business and legal environment, the telecommunications infrastructure, customer access, and business and financial service quality. The Singaporean government has actively sought to attract such offices since 1987, when it established tax incentives through the operational headquarters (OHQ) scheme, and the business headquarters (BHQ) scheme followed in 1994 for less influential regional offices. The evidence is very mixed, however, as to the extent to which these incentives actually influence location decisions, particularly with respect to US corporations. RHQs from US firms account for a significant proportion of the total: Perry *et al.* estimated that 34 per cent of manufacturing and 46 per cent of service regional offices were American, with US firms accounting for 60 per cent of offices with over 25 staff.

A key argument of this chapter is that we are seeing the functional integration of activities across national subsidiaries but within these regional structures. The mere presence of regional offices does not necessarily indicate such integration: in their simplest form, RHQs could just be consolidating national financial accounts, and acting as a conduit for managerial information from global headquarters within a broadly multidomestic structure. However, empirical evidence suggests that many of the regional offices based in Singapore undertake a range of other functions, including managing corporate expansion through the broader region, establishing both local and pan-regional alliances and partnerships, and providing a variety of technical support to affiliates (Perry, 1992; Perry, Yeung and Poon, 1998), thereby fulfilling what Lasserre (1996) describes as both an 'integrative' and an 'entrepreneurial' role. In this sense, the various national subsidiaries in Southeast Asia may depend on their RHQ for a wide variety of internal services, expertise and support. This pattern is also borne out in the sample of US IT firms being analysed here.

However, the evidence of functional integration goes beyond the role of the regional offices. Many of these offices are also operating as what might be called 'functional' offices (UNCTAD, 1993). As Table 5.1 has illustrated, perhaps the key functions performed by these firms in the region are the sales, marketing, implementation and support of a variety of software and service activities. Many of the skills needed for such operations are not present in each national territory, particularly in an emerging and rapidly developing regional structure. Rather, firms make use of regional pools of talent, in terms of both marketing and/or technical ability, and move individuals or teams of staff around as and when necessary. What is emerging, then, is a pattern of functional integration at the level of Southeast Asia or the Asia–Pacific: not in terms of products, but in terms of flows of people across a matrix form of organization loosely segmented by both geography and vertical market, and in some cases other skill groupings. Functional integration in service activities needs to be seen in terms of a dependence upon embodied skills available from other countries in the region. The manifestation of this integration is a fluid system of project-based teams of staff characterized by considerable mobility both between projects and between national territories within the region, depending on the particular patterns of demand for different kinds of expertise. Such processes are much harder to identify and characterize than the international functional integration of manufacturing activities that may be 'visible' through the adding of value to products as they make their way towards final assembly. Understanding the cross-border organization of service activities is critical, however, if the complex geographies of current globalization and regionalization processes are to be understood. To develop this argument, the next section moves on to consider some of the different dimensions of complexity that may occur with a multiregional structure.

# DIMENSIONS OF SPATIAL VARIATION

While diagrams such as Figure 5.1 are useful to delimit basic processes and broad dynamics of change, they can of course in no way convey the complexity of organizational structures that are found on the ground. In particular, such a schematic characterization cannot easily reveal the different levels of integration both across and within different regional structures. Equally, it is difficult to convey the 'nested' nature of regional structures whereby subregions may exist within the broad triadic regional structures. This section illustrates some different aspects of this complexity by considering three different dimensions of spatial variation – as they are termed here – that may occur within a broadly multiregional structure.

### Interregional Variations

Put simply, firms do things differently in different regions. There are two key aspects here. Firstly, there is variation simply in the presence of corporate activities across the different regions, and secondly, there is variation in the extent to which firms integrate activities between the different regions. As Dicken *et al.* (1997, p.163) argue,

> such regional systems vary greatly in their degree of self-containment, particularly because firms show a very high propensity to retain home-country control of the highest-order functions. They also vary in the extent to which their production system in one region is connected functionally to that in other regions, apart from the firm's home region.

Sun Microsystems can be used to illustrate this issue at a very basic level. Sun produces a range of hardware and software products. As with many such firms, Sun undertakes most of its research and development in North America, although there are small research centres in Europe (for example, Ireland) and the Asia–Pacific (for example, India) which together form an integrated network. Manufacturing is undertaken within both North America and Europe, but there is no operation as yet in the Asia–Pacific, to which products are simply exported. In terms of sales and marketing, and support functions, these are present and coordinated within all three regions. The role of Singapore within this structure is to provide the headquarters function (finance, human resources and so on) for the 'South Asia' sales and marketing operations (Southeast Asia and India), top-level technical support for the Asia–Pacific, and internal IT coordination and control for the Asia Pacific. Thus the picture is complicated by the integration of various functions at the level of subregion (in this case South Asia) when firms subdivide the Asia–Pacific region, thereby creating another possible scale of integration.

**Intraregional Variations**

The second dimension of variation is that different parts of an international business firm will or may exhibit different spatial strategies. Goodman (1997), for example, argued that corporate strategy cannot be considered as a monolithic whole, but rather processes of what he terms internationalization, multinationalization, transnationalization and globalization may coexist within the same firm. Similarly, in an analysis of Nestlé's operations in Southeast Asia, Pritchard and Fagan (1999) illustrate how the three different realms of production, realization and reproduction are characterized by different geographies and spatial strategies. In the case of US IT firms in Southeast Asia, while sales, marketing and implementation functions tend to be distributed and coordinated across a range of countries, manufacturing operations, where present, tend to be much more concentrated.

Figure 5.3 shows the example of the software product firm Microsoft, which in 1999 moved from a system of national production and fulfilment facilities to a consolidated operation for the Asia–Pacific in Singapore, thus bringing it in line with the Seattle and Ireland operations for the other two major regions. Microsoft continues to operate a range of national subsidiaries in the region to handle sales and marketing. The vast majority of research and development is concentrated in Seattle. The Asia–Pacific division of personal computer manufacturer Dell (which excludes Japan) provides another example. The Asia–Pacific headquarters is officially in Hong Kong, but most of the key functions are located in Singapore, including the new on-line data and development centre for managing on-line sales. In terms of manufacturing, Penang, Malaysia provides products for the whole Asia–Pacific with the exception of China and Hong Kong, which are supplied through a second plant in Southern China. There are support call centres in China (for China/Hong Kong), Australia (for Australia/New Zealand) and then Penang for the rest of the Asia–Pacific region. Financial consolidation and billing for the whole region is also undertaken in Penang, while there are sales and marketing operations in Australia, Brunei, China, Hong Kong, India, Japan, Korea, Macau, Malaysia, New Zealand, Singapore, Taiwan and Thailand. In addition, 44 distributors serve another 26 markets. Database software firm Oracle coordinates implementation, finance, basic training and personnel at the level of the national subsidiary in the Asia–Pacific, but centralizes others, such as advanced training, legal affairs and the marketing of new products, at the regional level. Even this,

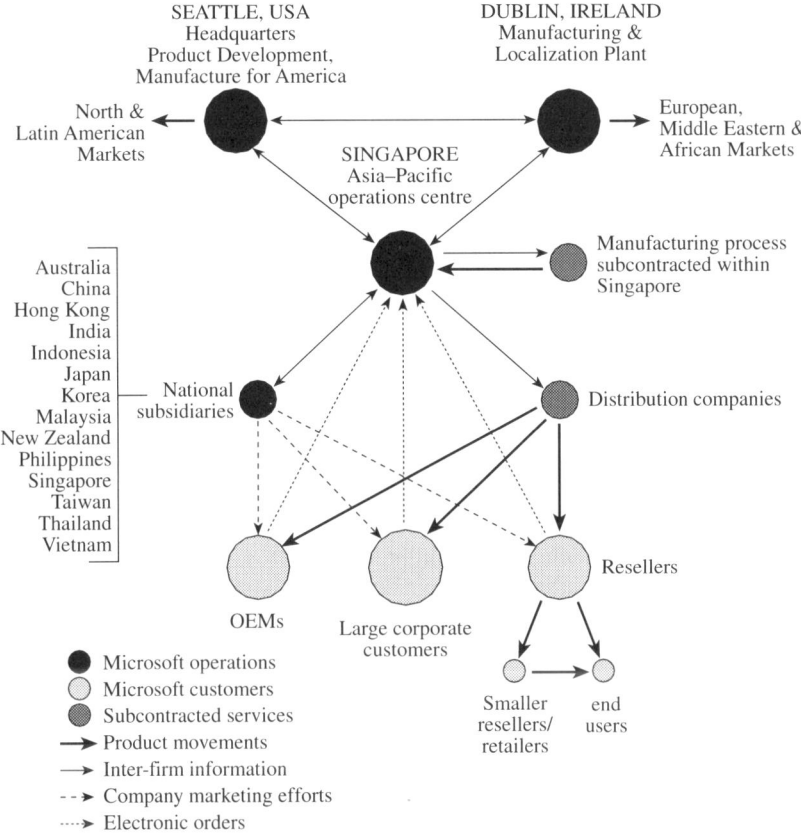

SEATTLE, USA
Headquarters
Product Development,
Manufacture for America

DUBLIN, IRELAND
Manufacturing &
Localization Plant

North &
Latin American
Markets

SINGAPORE
Asia–Pacific
operations centre

European,
Middle Eastern &
African Markets

Manufacturing process
subcontracted within
Singapore

Australia
China
Hong Kong
India
Indonesia
Japan
Korea
Malaysia
New Zealand
Philippines
Singapore
Taiwan
Thailand
Vietnam

National
subsidiaries

Distribution companies

OEMs

Large corporate
customers

Resellers

● Microsoft operations
○ Microsoft customers
◍ Subcontracted services
→ Product movements
→ Inter-firm information
--► Company marketing efforts
····► Electronic orders

Smaller
resellers/
retailers

end
users

*Source:* Coe (1999, p.48).

*Note:* OEM is original equipment manufacture. This entails a specific form of
subcontracting under which the local firm produces a good to the exact specification
of the foreign firm; the foreign firm then markets the product through its own
distribution channels, under it own brand name.

*Figure 5.3    Microsoft's global manufacturing and distribution operations*

however, is a gross simplification of a complex reality in which responsibilities for each activity are split – and subject to constant change – between
the different levels of organization. These brief examples serve to illustrate
a simple but important point, namely that very different geographies, or
scales of integration, characterize different types of activities within the
region. Moreover, these different scalar configurations may vary from
region to region, or be broadly the same, as in the Microsoft case.

**Intra-activity Variations**

The third dimension of variation is that a single activity can be undertaken in different ways within a region. Not only that, but, as has already been indicated, how a product or service is provided in a particular place may change over time. Hence it is important to consider both the temporal and spatial patterns of expansion for a particular activity within evolving regional structures. Many models of international business are concerned with modes of entry into national territories rather than spatial evolution of operations within regional formations. Many of the firms studied here have been doing business with customers in the major markets of North America and Europe for much longer than in the Asia–Pacific, the latter representing a more dynamic arena of international expansion through which to study these dynamics.

Considering the sales and marketing networks of the US IT firms provides a useful illustration. It is possible to build up a typical picture of expansion from the various sample firms. Many firms start by establishing a regional headquarters in Singapore and initially serve the region from there. They then establish relationships with partners and distributors across the region to provide access to key national markets. As business develops over time, the firm may buy shares in, or buy out completely, their existing distributors, while at the same time seeking other distributors in other emerging national markets. In this sense, Singapore is often used as an initial 'bridgehead' in the region from which to expand the network of offices, partners and distributors. This generalized pattern means there tends to be a level of regional integration from the outset, with functions being organized, monitored and controlled from Singapore.

The position of database software company Informix in 1999 freezes these dynamics at one point in time. In addition to sales and marketing staff for the Singapore market, Singapore is also home to the Asia–Pacific and 'Southeast Asia and India' headquarters, and product support for 'Southeast Asia and India'. Singapore also fits into a top-level 'follow-the-sun' global support system. In terms of sales and marketing, it has offices in India, Thailand, Malaysia and the Philippines where, in each case, distributors were subsequently acquired. The firm operates through distributors in Indonesia, Vietnam and Sri Lanka, and where necessary sells into the very small markets of Cambodia, Laos and Myanmar through ad hoc distributors, or partners such as the major systems integrators or management consultancies. The point here is that the same activity may be undertaken in different ways (or of course not at all) in different parts of the region.

## CONCLUSION

To conclude, let us briefly consider the main arguments of this chapter. Firstly, there is clearly a need for precision in the definition and use of key terms surrounding international business, such as regionalization, internationalization and globalization. Considering the extent to which economic activity is functionally integrated across different geographic scales provides a useful way forward. Secondly, it seems that the triadic region is an increasingly important level of strategic organization for TNCs with a global reach, not just for manufacturing operations, but also for a broad range of service activities. The term 'multiregional' firm, is used to describe a firm that integrates its various operations within these major regions but not – or to a lesser extent – across them. More work needs to be done on these regional levels of integration, as many models of international business are seemingly overly preoccupied with the national and global scales. Thirdly, and contra many homogenizing accounts of international business, heterogeneous and complex organizational structures can be found within broadly multiregional firms. Brief examples from US IT industry investments in Southeast Asia have only started to hint at this complexity.

The task for future research is to identify and study the overlapping scales of integration, be they national, subregional, regional or truly global, that characterize different parts of TNC activity. This is not merely an academic exercise, but one with profound implications for understanding the development dynamics of cities and countries that intersect these organizational structures in different ways. For example, Singapore has attracted significant numbers of high-quality jobs through being home to both the Southeast Asian or Asia–Pacific headquarters and support functions of many US IT firms, and substantial numbers of staff that form the basis of regional sales, marketing and implementation teams. While the software and services in question do not originate in Southeast Asia, Singapore has attracted a large number of skilled jobs involved in marshalling the dissemination of these services into the Asia–Pacific regional market place. Fourthly, future research work is particularly urgent with respect to the service sector activities that account for an ever-increasing proportion of FDI flows. How broad ranges of 'people-oriented' service tasks are being organized within and across countries and regions is relatively poorly understood. Globalization research quite simply needs to move beyond its current preoccupation with manufacturing and formal R&D activity.

Finally, what are the implications of this analysis for our understanding of innovation within internationally organized service activities? Empirically, it is clear that, with a few notable exceptions (IBM, HP and Sybase, who have development centres in Singapore), US IT firms are undertaking the majority of their formal R&D in their home country. In this sense, there is little intersection between processes of internationalization and innovation. All the interviewed firms were asked whether and how innovation could take place within their Asia–Pacific operations, and most described mechanisms by which new ideas could be fed back to research and technical teams in the USA. However, in almost all cases, this was operating at the level of fairly minor adjustments to products and services in order to adapt or localize them for the peculiarities of the different national markets within Asia. This overall pattern of high dependency upon the innovative capacity of the firm in the USA is reflective of the fact that many firms have moved into the region relatively recently, simply to tap markets, and are still expanding aggressively. Research into much longer-standing US IT industry investments in Western Europe is likely to reveal more developed international innovation networks within these TNCs. Conceptually, it may be useful to characterize innovative activity through the notion of scales of integration. Innovation, in the IT industry at least, is perhaps one activity that tends to a lesser extent to be organized through the multiregional structures described in this chapter. In the firms studied here, innovation was undertaken either almost exclusively in the home country or, in a very few cases, through a global network of research and development centres developing products and services for the firm as a whole. Firms were not developing ranges of services and products for distribution within just one regional structure. Building on the theme of this chapter, far more research is needed, however, into the complex ways in which innovation is organized across national territories in a wide range of service activities.

## NOTE

1. The research on which this chapter is based was funded by a National University of Singapore academic research grant (RP3982090).

## REFERENCES

Allen, J. (1992), 'Services and the U.K. space economy: regionalization and economic dislocation', *Transactions of the Institute of British Geographers,* **17** (3), 292–305.
Bartlett, C.A. and S. Ghoshal (1989), *Managing across borders: the transnational solution*, London: Century Business.
Bélis-Bergouignan, M-C., G. Bordenave and Y. Lung (2000), 'Global strategies in the automobile industry', *Regional Studies*, **34** (1), 41–53.

Coe, N.M. (1997), 'Internationalization, diversification and spatial restructuring in transnational computer service firms: case studies from the UK market', *Geoforum*, **28** (3–4), 253–70.

Coe, N.M. (1999), 'Emulating the Celtic Tiger? A comparison of the software industries of Singapore and Ireland', *Singapore Journal of Tropical Geography*, **20** (1), 36–55

Cox, K. (1998), 'Spaces of dependence, spaces of engagement and the politics of scale, or: looking for local politics', *Political Geography*, **17** (1), 1–23.

Daniels, J.D. (1987), 'Bridging national and global marketing strategies through regional operations', *International Marketing Review*, **4** (1), 29–44.

Daniels, P.W. (1993), *Service industries in the world economy*, Oxford: Blackwell.

Department of Commerce (2000), *Survey of Current Business*, September, Washington, DC: US Department of Commerce.

Dicken, P. (1998), *Global shift: transforming the world economy*, London: Paul Chapman.

Dicken, P. and C. Kirkpatrick (1991), 'Services-led development in ASEAN: transnational regional headquarters in Singapore', *The Pacific Review*, **4** (2), 174–84.

Dicken, P. and H.W-C. Yeung (1999), 'Investing in the future: East and Southeast Asian firms in the global economy', in K. Olds, P. Dicken, P.F. Kelly, L. Kong and H.W-C. Yeung (eds), *Globalisation and the Asia Pacific: contested territories*, London: Routledge, pp.107–28.

Dicken, P., J. Peck and A. Tickell (1997), 'Unpacking the global', in R. Lee and J. Wills (eds), *Geographies of Economies*, London: Edward Arnold, pp.158–66.

Dunning, J.H. (1993), *Multinational enterprises and the global economy*, Reading, MA: Addison-Wesley.

*The Economist* (1999), 'Last men standing', 3 April, pp.59–60.

Fagan, R. (1997), 'Local food/global food: globalization and local restructuring', in R. Lee and J. Wills (eds), *Geographies of Economies*, London: Edward Arnold, pp.197–208.

Forsgren, M., U. Holm and J. Johanson (1995), 'Division headquarters go abroad: a step in the internationalization of the multinational corporation', *Journal of Management Studies*, **32** (4), 475–91.

Goodman, D. (1997), 'World-scale processes and agro-food systems: critique and research needs', *Review of International Political Economy*, **4** (4), 663–87.

Greenwood, R., T. Rose, J.L. Brown., D.J. Cooper and B. Hinings (1999), 'The global management of professional services: the example of accounting', in S.R. Clegg, E. Ibarra-Colado and L. Bueno-Rodriquez (eds), *Global management: universal theories and local realities*, London: Sage, pp. 265–96.

Ho, K. (1998), 'Corporate regional functions in Asia Pacific', *Asia Pacific Viewpoint*, **39** (2), 179–91.

Howells, J. and M. Wood (1993), *The globalization of production and technology*, London: Belhaven.

Lasserre, P. (1996), 'Regional headquarters: the spearhead for Asia Pacific markets', *Long Range Planning*, **29** (1), 30–37.

Lévy, B. (1995), 'Globalization and regionalization: towards the shaping of a tripolar world economy?', *The International Executive*, **37** (4), 349–71.

Markusen, A. (1999), 'Fuzzy concepts, scanty evidence, policy distance: the case for rigour and policy relevance in critical regional studies', *Regional Studies*, **33** (9), 869–84.

Mathews, J.A. (1999), 'A silicon island of the East: creating a semiconductor industry in Singapore', *California Management Review,* **41** (2), 55–78.

Mirza, H. (ed.) (1998), *Global Competitive Strategies in the New World Economy: Multilateralism, Regionalization and the Transnational Firm*, Cheltenham, UK and Lyme, US: Edward Elgar.

Perry, M. (1992), 'Promoting corporate control in Singapore', *Regional Studies,* **26** (3), 289–94.

Perry, M., J. Poon and H. Yeung (1998), 'Regional offices in Singapore: spatial and strategic influences in the location of corporate control', *Review of Urban and Regional Development Studies*, **10** (1), 42–59.

Perry, M., H. Yeung and J. Poon (1998), 'Regional office mobility: the case of corporate control in Singapore and Hong Kong', *Geoforum*, **29** (3), 237–55.

Poon, J. (1997), 'The cosmopolitanization of trade regions: global trends and implications, 1965–1990', *Economic Geography*, **73** (4), 390–404.

Poon, J., E.R. Thompson and P.F. Kelly (2000), 'Myth of the triad? The geography of trade and investment blocs', *Transactions of the Institute of British Geographers,* **25** (4), 427–44.

Pritchard, B. (2000), 'Geographies of the firm and transnational agro-food corporations in East Asia', *Singapore Journal of Tropical Geography*, **21** (3), 246–62.

Pritchard, B. and R. Fagan (1999), 'Circuits of capital and transnational corporate spatial behaviour: Nestlé in Southeast Asia', *International Journal of the Sociology of Agriculture and Food*, **8** (1), 3–20.

Reich, R.B. (1991), *The work of nations*, New York: Vintage Books.

Rugman, A. (2000), *The end of globalization*, London: Random House.

Ruigrok, W. and R. van Tulder (1995), *The logic of international restructuring*, London: Routledge.

Smith, N. (1993), 'Homeless/global: scaling places', in J. Bird, B. Curtis, T. Putnam, G. Robertson and L. Tickner (eds), *Mapping the futures: local cultures, global change*, London: Routledge, pp.87–119.

Stallings, B. (ed.) (1995), *Global change, regional response: the new international context of development*, Cambridge: Cambridge University Press.

Swyngedouw, E. (1997), 'Neither global nor local: 'glocalization' and the politics of scale', in K.R. Cox (ed.), *Spaces of globalization: reasserting the power of the local*, New York: Guilford, pp.137–66.

UNCTAD (1993), *The world investment report, 1993: transnational corporations and integrated international production*, New York: United Nations.

UNCTAD (2001), *The world investment report, 2001: promoting linkages,* New York: United Nations.

Wong, P.K. (1998), 'Globalization of US–Japan production networks and the growth of Singapore's electronics industry', in Institute of Developing Economies (IDE) and Japan External Trade Organization (JETO) (eds), *Can Asia recover its vitality? Globalization and the roles of Japanese and US corporations*, Tokyo: IDE, pp.89–105.

Yeung, H., J. Poon and M. Perry (2001), 'Towards a regional strategy: the role of regional headquarters of foreign firms in Singapore', *Urban Studies*, **38** (1), 157–83.

# 6. National versus international effects in regional concentration of European innovative business services

**Luis Rubalcaba-Bermejo and David Gago-Saldaña**

## INTRODUCTION

This chapter focuses on the role of international and national effects in regional concentration of innovative services.[1] The increasing influence exerted by globalization on service industries is epitomized by the case of innovative advanced services, many of which were traditionally located only according to national, regional or urban patterns. The process of globalization might suggest a certain uneven balance between national and international patterns. This chapter aims to present some European comparative results through the use of some key explanatory factors for business service location (qualifications, innovative performance, density and economic development). The chapter discusses some of the driving forces influencing locational patterns both at a national and at an international level, with a special emphasis on innovative (advanced) services.

The analysis is based on unpublished EUROSTAT data on the number of business services employees at NUTS1 or/and NUTS2 levels (Nomenclature of Territorial Units for Statistics) and considers three main categories of business services: advanced, traditional and operational, even though two other categories, namely computer and research and development services (R&D), have also been included selectively in some analyses. These latter two categories come from a further desegregation of the advanced business services.[2] Only the regions of five countries make up the European comparative results presented: France, United Kingdom, Austria, Belgium and Finland. The data refer to the year 1997.

The chapter is organized as follows: the first section sets out an introductory explanation of the links between business services, globalization and regional development, concluding with the exposition of the main hypothesis discussed in the chapter. The second section highlights some empirical descriptive evidence concerning the importance of business services (especially advanced business services) for the selected regions as a leading to the third section, which contains the main results of the chapter. The section explores the importance of some of the locational factors likely to affect business services deployment carrying out an econometric analysis. Moreover, the validity of the locational factors is studied after controlling for differences in national systems ('national effects') and the role of capital regions ('international effects'). The fourth section is devoted to a study of the particular relationship between business services and GDP per capita in more depth, assessing the special role of capital regions. Finally, some conclusions and closing remarks are provided in the fifth section.

## INNOVATIVE BUSINESS SERVICES IN THE GLOBAL ECONOMY

The first modern wave of globalization between 1870 and 1914 was characterized by the 'industrialization' process (the international consolidation of manufacturing factories born in the Industrial Revolution) and the growth of international firms (those trading with several countries). The second wave of modern globalization that took place after the 1960s had, amongst others, two peculiarities, which helped reinforce the role of services in dynamic economic change: the consolidation of the service economy and the growth of transnational firms and networks.

The structural change described as the shift from manufacturing to services in developed countries is a complex process explained by many supply factors (for example, differences in productivity), demand factors (for example, Engel's law) and contextual factors (for example, market integration). Innovative business services have been very active in this process through, at least four different channels: in the first place, by providing new products in order to create new intermediate value added to any manufacturing or service production process (for example, design or quality control services); in the second place, by enabling a rational and safe international expansion (as in the case of international strategic consultancy services); thirdly, by creating new employment opportunities in highly labour-intensive services (where the productivity gains can be measured more in quality gains than in scale economies or costs gains); and fourthly, by promoting value added to the new technological tools supporting the globalization processes (computer or telecommunication services). The recent boom of the new economy and Internet has also been linked to those innovative services behind Internet hosts and providers, new telecommunication services, new marketing areas or new R&D-related activities. In a sense, the new economy can be defined as new technologies implemented by new innovative services. In fact, any Internet operation using both a mobile personal computer (PC) and a phone (no matter the level of sophistication) requires a previous set of innovative services linked to the design, production, marketing, implementation and use of hardware and software technology.

The second major feature of globalization, over which business services have been very active, can be identified as the gradual transition that international firms have experienced in the way of becoming multinational firms or networks. The new service economy has been closely linked to the trends towards flexible productive systems, which are the foundations of the international division of labour, the new possibilities of teleworking and new managerial practices for international business. Routine business services have allowed enterprises to concentrate on the core areas, while innovative business services have increased their strategic capabilities and competitive edge. Furthermore, business services become key tools for those enterprises that decide to move into foreign markets. In this sense, they require a wide number of innovative services, either to be advised (management consultancy), to test local markets (fairs and exhibitions, market research) or to implement the final decisions (legal services, strategic consultancy). Finally, many business services promote multinational firms directly by providing global services and by their own globalization processes.[3]

In the context of globalization, a service is considered innovative when it creates an incremental or radical change in dynamic production processes, products and organizational capabilities. Thus innovation is based on new knowledge and new resources applied to processes, products and organizations. Business services are innovative when they affect, in this sense, any of the key areas of an enterprise. Previous work by one of the authors (Rubalcaba-Bermejo, 1999) distinguishes five types of innovative functions related to business services: ICT technological, organizational, strategic, commercial and operational innovation functions. All these innovation functions can substantially increase organizational capabilities and knowledge of productive processes or products.

Along with the contribution of innovative business services to the globalization processes, business services themselves are strongly affected by global trends. A number of statistics can be used to exemplify this. In fact, business services, which represent around 10 per cent of value added and employment of European economies (excluding real state, renting and telecommunication services), cover 16 per cent of total European foreign direct investment (FDI) stocks (inward and outward) in the world (13 per cent outward; 19 per cent inward, inward FDI being 21 per cent larger than outward). In mergers and acquisitions (M&A), business services represent more modest economic values (7 per cent of the total sales and 4 per cent of total purchases). Of course, figures are much less prominent when the scope of analysis is international trade, due to the reduced role of services in total trade (19 per cent of world trade in 1999, and 5 per cent due solely to business services). However, in any case, business services have been the most active service sector, according to many globalization indicators (Rubalcaba-Bermejo and Cuadrado-Roura, 2002). They report an average annual growth of 17 per cent in European international trade between 1989 and 1998 and 33 per cent between 1992 and 1998. Another prominent role of business services applies to FDI, especially in European outward FDI between 1992 and 1998 (32 per cent). In M&A, the leading service sectors in the 1990s were transport and communication services, hotels and restaurants and financial services. On the other hand, business services have reported important annual growth rates (54 per cent), usually pioneering the number of M&A operations (European Economy, 2000). In summary, business services are major contributors to service globalization growth (except in M&A) in absolute terms, even though some other smaller sectors seem to be more global and active in relative terms (according to relative size), especially transport and communication services and, overall, financial services.

The reasons why business services are so active in the globalization processes are manifold. For example, the theories explaining the advantages of becoming multinational (the OLI model or the eclectic paradigm – see Dunning, 1993) can be at least partially valid for understanding service globalization, even if purely economic factors have less importance (Petit, 1986; Riddle, 1986) and should be reformulated according to the firm's strategy and country resources and endowments (Roberts, 1998). In the case of innovative business services, locational, strategic and cultural factors are crucial in explaining the international growth of some relevant management consultancy firms and other advanced business service firms after the Second World War, which often became global owing to 'following the clients' and 'following the leaders' effects.

## IMPLICATIONS FOR REGIONAL DEVELOPMENT

The role of business services in regional development has been widely explored and recognized elsewhere (by way of example, Daniels and Moulaert, 1991; Daniels *et al.*, 1993; Daniels, 1993; De Bandt and Gadrey, 1994; Bonamy and Valeyre, 1994; Marshall and Wood, 1995; Illeris, 1996). Business services are primarily concentrated in the most dynamic areas, basically corresponding to a few metropolitan regions. They usually act not only as followers of the local economic base, but also as poles of attraction of investment and as sources of agglomeration economies. In any case, the contribution of the different categories of business services is quite disparate. For example, some ICT or R&D-related services can furnish certain decentralization processes since proximity requirements are not so important as in traditional services (Illeris, 1994). However, most advanced innovative business services contribute to further centralize economic activity. In fact, many of the most innovative and advanced services are overwhelmingly concentrated at the higher levels of the regional or urban hierarchies and, even within a given city, at the most prestigious places and districts. The role of ICT in decentralizing innovative services is still rather modest, although new opportunities are emerging, even in rural areas (OECD, 2001).

The context of globalization pushes business services to be more competitive, and the decision of where to locate becomes crucial in order to gain competitive advantage. In this sense, two types of geographic spaces are the most favoured in the new global arena: safe central places where the leading firms in the sectors are already operating and hosting current and potential clients, and high standing places where working conditions are excellent and services can be provided at a distance.[4]

It is worth mentioning that knowledge is so embedded in many innovative services that an international division of labour is difficult to operate. Most transnational business service firms are likely to replicate the service content with more or fewer similarities with respect to the service provided at the headquarters. The prestige associated with image and brand, and a certain methodology and organization remain the same or similar across the international local offices, but only a reduced and very specialized number of services are provided worldwide as in a real global firm. Expectation, reputation and prestige play a decisive role (Aharoni, 2000) as do the creation, accumulation, transfer and protection of knowledge (Gross, 2000). All of this restricts the service standardization processes up to a point, and forces most business services to remain local or to be provided on a local basis, even if they are international. Business services, and innovative business services in particular, are very interactive and knowledge-intensive, and expectations are high to keep a fluent coproduction between the client and the provider. This effect may not be achieved if a wrong location is chosen, if local business conditions decline or if the service is not provided in the right interactive way (the service may be perceived as too standard or too easily acquired by the competitors). Business service 'internationalization' presupposes 'nationalization', in other words, adaptation to the regulatory, economic, social and cultural parameters of the country or region, or market where the clients operate (Rubalcaba-Bermejo, 1999).

The strong links between business services and local development lead us to hypothesize the existence of a strong 'national effect' in their concentration across Europe, since artificial and natural barriers still remain very strong (European Commission, 2001), and multinational firms have to replicate similar services in each country. Different economies and development degrees, different legal systems, different languages, different cultures and different local bases are all ingredients that lead us to expect a strong national effect, as the recent work edited by Wood (2001) shows when analysing the relationships between consultancy services and innovation.

However, at the same time, globalization is increasingly opening up possibilities for revamped locational trends to flourish in certain business services. The main consequence is the reinforcing role of business services in the upper urban hierarchy: the main international villages should concentrate the most advanced and innovative business services, no matter what the country where they are located. Amongst other possible patterns, this is the hypothesis to be tested in the next section and labelled the 'international effect', in possible opposition to the 'national effect'. In principle, the impact of globalization might suggest a certain shift from the predominance of the national to the international effect. Moreover, the extent of both effects must be assessed in the context of a comprehensive analysis that includes other major locational factors. These range from some traditional locational factors (GDP per capita, the economic base or urban density, amongst others) to explanations based more on intangible regional competitive advantages, such as qualifications or innovative performance. Such is the purpose of the fourth section, once some preliminary data on business services in European regions have been presented in the next section.

## BUSINESS SERVICES IN EUROPEAN REGIONS: A PRELIMINARY EMPIRICAL OVERVIEW

Table 6.1 presents the relative weight of business services at NUTS1 level from France, the UK, Belgium, Austria and Finland. Regions are ranked following two criteria (labelled A and B, respectively), namely the number of employees in business services per 1000 inhabitants and per 1000 persons employed, even though the ranking turns out to be quite similar. The UK is the country reporting the highest figures, with 50 employees per 1000 inhabitants and 111 per 1000 persons employed. Belgium and France follow then with 30 (82), and 28 (75) employees, respectively. Finland has 24 (59) employees, and Austria 18 (41). These differences amongst countries reported under both criteria are mainly explained by the leading role of certain regions. For example, the clear UK leadership is due to the economic weight of the South East, (London included) and the relative high development of most UK regions. In this sense, the West Midlands, East Anglia, the North West, the East Midlands, the South West and Yorkshire all rank between five and 10 in this regional hierarchy of business services.

This picture illustrates a preliminary interesting point made in the last section: the differences in this European hierarchy of NUTS1 regions are given not only by the role of European capitals, but also by the presence of a national component. London, Paris or Brussels concentrate the highest number of business services per 1000 inhabitants or persons employed, at the same time reflecting the relative situation of the countries they belong to. On the other hand, the Vienna region has only around 37 (80) employees per 1000 inhabitants (persons employed). Vienna is a second-tier European city, but is leader in a country where the development of business services is shaped by a peculiar industrial structure. In fact, Austria's industrial model (a reflection of the German model), widely promotes the production of in-house business services, whereas free-standing firms are less developed. Conversely, the Anglo-Saxon model led by the UK shows a different pattern of business services production, externalization being a common feature, whereby business services firms play a more important role. In addition, the UK economy is much more business services-oriented; in this sense, different European studies on business services agree on this supremacy of UK business service activities (see Rubalcaba-Bermejo, 1999), as markets are more mature and multinationals are more widely present. Most European multinationals in the sector are British, some are Dutch, and relatively few, for the relative country's size, are French or German. Of course, there are many from the United States.

The distribution of advanced business services differs to some extent from the distribution of the business services as an aggregate. In this case, the hierarchy is slightly different, the pole position being represented now by the South East region if the first criterion is used (26) or Brussels (71) using the second one. Generally speaking, the UK regions are again ranked at the top, but this is more evident using the first and not the second criterion; in fact, according to the second criterion, two Belgian regions, namely Walloon (31) and Vlaams Gewest (38), report higher figures than that observed in most British regions. The development of advanced business services in most Belgian regions can be partly explained by the impressive development of some of these activities in the Netherlands (as some of them can be setting some affiliates in Belgium), the role of the international institutions (and the EU, the activity of which promotes many supporting consultancy services) and high levels of economic activity and income (Belgium is the first of the five countries in terms of GDP per capita).

*Table 6.1  The importance of business services in European regions: breakdown by type of business services (five countries)*

| | Ranking | | Total | | Advanced | | Traditional | | Operational | | R&D | | Computer | |
|---|---|---|---|---|---|---|---|---|---|---|---|---|---|---|
| | A | B | A | B | A | B | A | B | A | B | A | B | A | B |
| Reg. Brussels-cap. | 1 | 1 | 82 | 241 | 24 | 71 | 36 | 106 | 21 | 63 | 2 | 5 | 6 | 19 |
| South East | 2 | 2 | 75 | 161 | 26 | 56 | 30 | 65 | 19 | 41 | 3 | 7 | 8 | 18 |
| Ile de France | 3 | 3 | 71 | 166 | 15 | 35 | 37 | 87 | 19 | 45 | 1 | 2 | 10 | 24 |
| **United Kingdom** | **4** | **4** | **50** | **111** | **16** | **35** | **19** | **43** | **15** | **37** | **2** | **4** | **4** | **10** |
| West Midlands | 5 | 5 | 45 | 102 | 15 | 34 | 15 | 34 | 15 | 35 | 0.8 | 2 | 3 | 7 |
| East Anglia | 6 | 7 | 40 | 86 | 14 | 29 | 15 | 33 | 11 | 24 | 3 | 6 | 3 | 7 |
| North West | 7 | 6 | 40 | 94 | 12 | 27 | 16 | 37 | 13 | 30 | 0.6 | 2 | 3 | 6 |
| East Midlands | 8 | 9 | 39 | 85 | 13 | 29 | 14 | 30 | 12 | 26 | 0.9 | 2 | 3 | 6 |
| South West | 9 | 8 | 39 | 85 | 12 | 27 | 15 | 32 | 12 | 25 | 0.9 | 2 | 3 | 7 |
| Yorkshire and Humberside | 10 | 10 | 36 | 82 | 9 | 21 | 13 | 29 | 14 | 32 | 0.5 | 2 | 2 | 5 |
| **Belgium** | **11** | **11** | **30** | **82** | **14** | **39** | **9** | **25** | **7** | **18** | **1** | **3** | **2** | **6** |
| Scotland | 12 | 15 | 30 | 69 | 9 | 21 | 8 | 19 | 13 | 30 | 0.9 | 2 | 2 | 4 |
| **France** | **13** | **13** | **28** | **75** | **5** | **13** | **14** | **38** | **9** | **24** | **0.5** | **1** | **3** | **8** |
| Vlaams Gewest | 14 | 14 | 28 | 72 | 15 | 38 | 8 | 20 | 5 | 14 | 0.8 | 2 | 2 | 6 |
| North | 15 | 16 | 28 | 68 | 6 | 14 | 11 | 27 | 11 | 26 | 0.2 | 0.5 | 1 | 3 |
| Méditerranée | 16 | 12 | 25 | 76 | 4 | 11 | 11 | 35 | 10 | 30 | 0.5 | 2 | 2 | 6 |
| Manner-Suomi | 17 | 18 | 24 | 59 | 5 | 14 | 10 | 26 | 8 | 20 | 0.7 | 2 | 3 | 8 |

| | | | | | | | | | | | | | |
|---|---|---|---|---|---|---|---|---|---|---|---|---|---|
| **Finland** | **18** | **19** | **24** | **59** | **5** | **13** | **10** | **26** | **8** | **20** | **0.7** | **2** | **3** | **8** |
| Wales | 19 | 21 | 22 | 55 | 5 | 11 | 8 | 21 | 9 | 23 | 0.1 | 0.2 | 1 | 2 |
| East Austria | 20 | 26 | 22 | 49 | 6 | 13 | 8 | 18 | 8 | 18 | 0.1 | 0.3 | 3 | 6 |
| Nord–Pas-de-Calais | 21 | 17 | 21 | 67 | 2 | 8 | 11 | 37 | 7 | 22 | 0.1 | 0.3 | 2 | 5 |
| Sud-Ouest | 22 | 22 | 20 | 53 | 4 | 10 | 10 | 25 | 7 | 17 | 0.9 | 2 | 2 | 6 |
| Est | 23 | 23 | 20 | 53 | 3 | 7 | 9 | 25 | 8 | 21 | 0.4 | 1 | 2 | 4 |
| Northern Ireland | 24 | 25 | 19 | 50 | 3 | 8 | 8 | 22 | 8 | 21 | 0.3 | 1 | 1 | 2 |
| Walloon | 25 | 20 | 19 | 57 | 10 | 31 | 4 | 13 | 4 | 13 | 1 | 4 | 1 | 2 |
| Bassin parisien | 26 | 24 | 19 | 52 | 3 | 7 | 9 | 25 | 7 | 20 | 0.4 | 1 | 1 | 3 |
| **Austria** | **27** | **27** | **18** | **41** | **5** | **11** | **6** | **14** | **7** | **16** | **0.1** | **0.3** | **2** | **4** |
| West Austria | 28 | 29 | 17 | 37 | 5 | 10 | 5 | 11 | 7 | 16 | 0.0 | 0.1 | 1 | 3 |
| Ouest | 29 | 28 | 15 | 40 | 2 | 6 | 8 | 21 | 5 | 13 | 0.1 | 0.5 | 1 | 4 |
| South Austria | 30 | 30 | 13 | 30 | 4 | 10 | 4 | 9 | 5 | 12 | 0.2 | 0.5 | 1 | 2 |
| Ahvenanmaavaland/åland | 31 | 32 | 10 | 20 | 2 | 4 | 5 | 10 | 3 | 7 | 0.2 | 0.3 | 2 | 3 |
| Centre-Est | 32 | 31 | 9 | 22 | 2 | 4 | 4 | 11 | 3 | 7 | 0.5 | 1 | 1 | 2 |

*Notes:*
[a] business services per 1000 inhabitants.
[b] business services per 1000 employed people. (General Industrial Classification of Economic Activities of the European Community (NACE) equivalence). Advanced = (sector 72 + sector 73 + sector 74.3 + sector 74.5); Traditional = (sector 74.1 + sector 74.2 + sector 74.4 + sector 74.8); Operational = (70 + 71 + 74.6 + 74.7); R&D = (73); Computer services = (72).

As far as traditional services are concerned, distribution of business services is more balanced, especially from the perspective of the first criterion. Accordingly, most regions report between eight and 15 employees per 1000 inhabitants. Only four regions are above 15: North West (16), Brussels (36), South East (30) and the leader, de France (37). Only a few regions (5) have under five employees: Westösterreich (5), Äland (5), Walloon (4), Südösterreich (4) and Centre-Est (4). The leader country is again the UK (19), followed by France (14), Finland (10) and Belgium (9).

Operational services have the most linear distribution from the European regions included in the sample. Deviations from the average are negligible, and the regional hierarchy is very similar to the global one taking the first criterion as a reference, whereas some minor differences are appreciable from the second criterion.

The main conclusion when analysing the distribution of computer services in these European regions is related to the fact that, contrary to the results reported for the operational services, the distribution is quite uneven. In this sense, a great deal of disparity between the three regions assuming a leading role (Ile de France, South East and Brussels) and the rest is reported. Finally, R&D services yield the most striking results, as they seem to present differences from the regularities depicted in the other categories of business services. By way of example, Ile de France, one of the leading regions, as seen above, reports a figure which is only half that of Walloon (ranked at the bottom in the global ranking), according to the second criterion. This preliminary evidence leads us to conclude that, despite possible statistical effects in the way of collecting data from specialized R&D firms, locational trends in R&D are not so predictable as one could hypothesize from looking at the other business services categories. This conclusion will be reinforced later on.

**Explanatory Factors for European Business Services**

The motivation of the following section is twofold. On the one hand, it aims to explore empirically some of the likely driving factors of business services concentration, which were briefly presented in the first section. On the other hand, we are interested in exploring whether the so-called 'national effect' detected in the former section does exist or, on the contrary, the 'international effect' is the significant one. Furthermore, the analysis will enable us to shed some light on the likely influence that both the 'national' and the 'international' effects have on these driving factors.

In order to explore these issues in detail, a series of regression analyses have been undertaken as shown in Tables 6.2 and 6.3. A first regression has been calculated to account for the 'national effect', which is done by

*Table 6.2   Explanatory factors by breakdown of business services categories: national v. international effects (standardized coefficients)*

| Dependent variable | Business services per Inhabitant | | | | | | | |
|---|---|---|---|---|---|---|---|---|
| | Total | | Advanced | | Traditional | | Operational | |
| Independent variables | National | International | National | International | National | International | National | International |
| Services | 0.17* | 0.16 | 0.09 | 0.11 | 0.19* | 0.13 | 0.27** | 0.23‡ |
| Patents | 0.24** | 0.13 | 0.28** | 0.19* | 0.19* | 0.07 | 0.19‡ | 0.08 |
| Qualifications | 0.04 | 0.28** | 0.09 | 0.48** | 0.09 | 0.19* | -0.13 | 0.05 |
| Density | 0.39** | 0.34** | 0.22* | 0.41** | 0.41** | 0.19‡ | 0.47** | 0.37** |
| GDP per capita | 0.26** | 0.10 | 0.23* | 0.28* | 0.29** | 0.03 | 0.16 | -0.05 |
| D1 | -23** | | 0.03 | | -0.29** | | -0.41** | |
| D2 | -0.17* | | -0.4** | | 0.07 | | -0.17* | |
| D3 | -0.39** | | -0.37** | | -0.3** | | -0.43** | |
| D4 | -0.14* | | -0.27** | | -0.02 | | -0.1 | |
| D5 | | 0.14 | | -0.32* | | 0.44** | | 0.23 |
| Corrected R² | 0.75 | 0.66 | 0.73 | 0.66 | 0.72 | 0.62 | 0.68 | 0.47 |
| Number of observations | 75 | 75 | 75 | 75 | 75 | 75 | 75 | 75 |

*Notes:*

‡ = coefficients significant at the 10% level,

† = coefficients significant at the 5% level,

* = coefficients significant at the 1% level (NACE equivalence); Advanced = (sector 72 + sector 73 + sector 74.3 + sector 74.5); Traditional = (sector 74.1 + sector 74.2 + sector 74.4 + sector 74.8); Operational = (70 + 71 + 74.6 + 74.7).

Table 6.3   Explanatory factors for advanced business services: national vs. international effects (standardised coefficients)

| Dependent variable | Business services per inhabitant | | | | | |
| --- | --- | --- | --- | --- | --- | --- |
| | Advanced | | Computer | | R&D | |
| Independent variables | National | International | National | International | National | International |
| Services | 0.09 | 0.11 | 0.3 | 0.24* | 0.2 | 0.29* |
| Patents | 0.28** | 0.19* | 0.34** | 0.31** | 0.37** | 0.35** |
| Qualifications | 0.09 | 0.48** | 0.14 | 0.14 | 0.33 | 0.34** |
| Density | 0.22* | 0.41** | 0.07 | −0.06 | −0.12 | 0.02 |
| GDP *per capita* | 0.23* | 0.28* | 0.29** | 0.11 | 0.13 | 0.13 |
| D1 | 0.03 | | −0.29** | | −0.18‡ | |
| D2 | −0.4** | | −0.03 | | −0.17 | |
| D3 | −0.37** | | −0.15 | | −0.13 | |
| D4 | −0.27** | | 0.03 | | −0.17 | |
| D5 | | −0.32* | | 0.31** | | −0.3‡ |
| Corrected R² | 0.73 | 0.66 | 0.63 | 0.57 | 0.45 | 0.45 |
| Number of observations | 75 | 75 | 75 | 75 | 75 | 75 |

*Notes:*
‡ = coefficients significant at the 10% level,
* = coefficients significant at the 5% level,
** = coefficients significant at the 1% level. (NACE equivalence); Advanced = (72 + 73 + 74.3 + 74.5); Computer services = (72); R&D = (73).

creating four dummy variables representing the five countries and taking the British case as a reference point to avoid the 'dummy trap'.[5] On the other hand, a second regression has been estimated to control for the 'international effect'. The latter is done through the creation of a dummy variable reporting the value one, if the region is the capital of the country, or zero, if not. Therefore this criterion coherently assumes that capital regions are the most exposed to international competition, and so are used as the framework to build the dichotomic variable.

Both regressions have been undertaken for the aggregate category of business services, in the first place, and for the five categories of business services, in the second place, as it is quite likely that these explanatory factors are not the same in every category. Preservation of the degrees of freedom has led us to use only five continuous independent variables in both the 'national' and 'international effects' regressions. The continuous independent variables have been the following: the share of services in the regional economy, an index of density of economic activity (calculated as the number of inhabitants per square kilometre), GDP per capita, and two indicators aiming to reflect regional competitive advantage, namely an index of qualifications (calculated as the share of population holding a university degree) and the number of patents per million inhabitants. The estimation has proceeded using ordinary least squares (OLS). All data refer to the year 1997.

The fit of the 'national effect' regression for the aggregate is better than that of the 'international effect' (0.75 versus 0.66), and the same applies for each of the five categories of business services. Nevertheless, substantial differences in the fit of the regressions are found amongst categories: advanced services where the fit is better (0.73 and 0.66 respectively) and R&D the poorest. The low fit observed in the latter case confirms that location trends in R&D seem to be partially governed by other factors not directly derived from the very regional conditions and competitive advantages (for example, public regional development policies), although statistical underrepresentation of the sector in certain areas can produce an unreal bias.[6]

Looking at the results, it would seem clear a priori that the so-called 'national effect' outstrips the 'international effect', as most dummies become significant (and with the expected sign) for the first effect, whereas the dummy representing the 'international effects' is not significant. In any case, this result must be interpreted with caution, since it is quite likely that these aggregate figures are masking different patterns and trajectories that are present when analysing particular categories of business services.

Some interesting results do flourish when looking at the coefficients reported in both regressions, illustrating the influence of the two approaches on the relevant variables included. In the first place, it is striking that the only variable that turns out to be significant in both cases (even at 1 per cent) is density, confirming the importance attached to agglomeration and concentration processes on business services location. On the other hand, patents is highly significant under the 'national effects' regression, but no longer significant under 'international effects'. On the contrary, qualifications is not significant under the first scenario, but highly significant when controlling for internationalization. This conclusion may be partly explained by the way that this indicator has been constructed. As explained above, the share of population holding a university degree has been used as a proxy of qualifications, and no further information has been taken into account. As expected, the coefficient is not statistically significant when controlling for country-specific effects, since educational levels are somewhat similar amongst regions pertaining to the same country. Conversely, the statistical significance of qualifications in the second regression illustrates the differences in educational levels from regions pertaining to different countries.

Patents represent the opposite case, as it turns out to be significant in the 'national effect' regression, but not in the 'international' one. The answer is in the line of the previous argument, but expressed the other way round: the number of patents is quite disparate amongst regions pertaining to the same country, because they reflect purely milieu conditions. Such an innovative environment can be stimulated or boosted through political intervention, but is somewhat related to the historical heritage and the institutional framework of the region. The significant and positive coefficient of patents in the first regression confirms the substantial differences in the interregional innovative performance.

Finally, the share of services in GDP and GDP per capita are only significant in the first regression, not in the second one, confirming that part of the link between both variables and business services endowment comes from a third effect, namely whether the region is capital or not.

When turning our attention to different categories of business services, it is worth mentioning that density and patents are the variables reporting the most significant results, patents being more important in advanced services (computer services and R&D included), and density in operational and advanced services other than computer and R&D. The latter result, as far as R&D is concerned, has to do with the fact that some R&D services are provided or supported by governments as a strategy to diffuse fully the benefits of technological development and innovation throughout the entire territory. Accordingly, private incentives governing location (of which density is one of the major factors) might not be the only force to consider. In the case of computer services, the lack of explanatory power of the density variable may be reflecting the loosening of spatial constraints brought about by the information and communication technologies (ICT) revolution, as some authors have emphasized (see, for example, Marshall and Wood, 1995).

On the other hand, the share of services in the regional economy reports the expected sign and turns out to be significant in some service categories, but not in others. This conclusion seems to indicate that a high presence of services in a region is not necessarily a natural precondition for the development of business services to take place in all cases, since the service sector comprises a great heterogeneity of activities, very often different in nature and condition. In relation to this, it is worth emphasizing the positive and significant effect of the 'services' variable on computer services deployment, highlighting the very well documented role of services as the main users of ICTs (see, for example, Gago-Saldaña, 2001).

The results reported by the dummy variables confirm the intuition expressed above, namely that the aggregate figures cannot be extrapolated when looking at different categories of business services. In that case, both 'national' and 'international' effects become significant in all categories (except for the operational services case). In the light of these results, one must conclude that the presence of a 'national effect' shaping the development of business services in every country does not prevent the 'international effect' from playing a role, so that what is 'national' and what is 'international' must be interpreted together to offer a thorough explanation of business services profiles in European regions.

Traditional services are the category with the most evident international orientation, this conclusion being justified on empirical grounds; in fact, most traditional services as here defined (for example, management consultancy, engineering, legal and economic consultancy, advertising) are run by multinational corporations. On the contrary, the negative sign reported for advanced services in the 'international effects' dummy would imply that this type of services is more developed in the non-capital regions. This, in principle, discouraging result may be reconciled by considering that R&D services (NACE 73) and the related tests and technical services (NACE 74·3) are responsible for such a conclusion. Conversely, computer services, which are a major activity within the advanced services category, report significant and positive coefficients, this result meaning that the development of computer services is one of the vectors enabling interdependence and hence pushing towards globalization.

# REGIONAL PROFILE OF EUROPEAN BUSINESS SERVICES ACCORDING TO PER CAPITA INCOME

GDP per capita has often been quoted as one of the most decisive factors in explaining business services location. Nevertheless, the previous analysis has proved that some factors other than GDP seem to be more pre-eminent, which poses an interesting question regarding the validity of the argument. In order to allow a better identification of the relationships between business services (per 1000 persons employed) and GDP per capita, a set of scatter plots has been built considering each of the five categories of business services, even though only the aggregate category is presented here. The results are shown for both the 'all regions' case (including NUTS1 and NUTS2) and the NUTS1 case only, in every category. This double perspective precludes some likely problems of interpretation at the NUTS2 level, where, as is usually the case in some urban regions, business services endowment is not coincident with the region where the employees live (that is, showing a phenomenon of daily commuting). Two main conclusions are worth mentioning when interpreting the relationship, considering business services in the aggregate (see Figure 6.1). First of all, the higher the income is, the more dispersion the values report in the case of all regions. In the second place, the analysis for the NUTS1 regions shows how dependent the correlation coefficient is on just two regions: Ile de France and Brussels. In fact, if these two regions did not exist, the correlation would even be negative. This conclusion means that, at low or medium levels of income per capita, the presence of business services varies a great deal and there is no clear correlation. At the same time, the result highlights again the importance of capital regions (the regions most exposed to international competition) as spaces favoured by business services location. Coming down into a second level of the urban hierarchy, things are not so obvious at the European level, even though they can still be valid for certain individual countries. It is important to note that the sample analysed includes the leader regions of UK, which in many cases are not leaders in terms of GDP per capita, so that a certain bias coming from the UK regions is found.

In the 'all regions' case, most positive deviations from the standard regression are due to the UK regions, Greater London having the highest deviation. Likewise, Berkshire, South East and Bedfordshire all have high deviations: much more significant business services than GDP per capita. On the contrary, the Vienna region depicts the opposite situation, with much less significant business services than income. Finally, the two Finnish regions represent opposite cases: Uusimaa reports high business services development, whereas Äland is clearly below the European average.

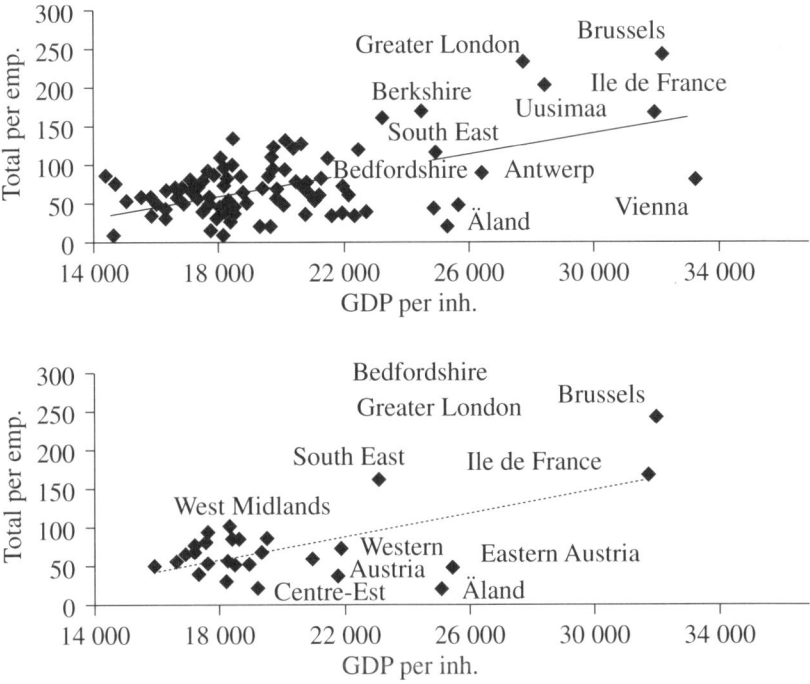

*Figure 6.1    Business services and GDP per inhabitant: all regions and NUTS 1*

As far as the NUTS1 analysis is concerned, the picture draws similar features. UK regions are overrated in terms of business services (South East and West Midlands are cases in point), as well as Brussels and Ile de France. Austrian regions (Eastern Austria and Western Austria), Åland and Centre–Est depict the opposite situation.

Finally, it is worth mentioning whether these conclusions are still valid when looking at every business services category. As in the aggregate case, the observations are more dispersed as higher levels of GDP per capita are reached, this conclusion being less evident in operational services than in any other case. The British regions are outstanding examples of positive deviations, showing higher business services development than expected according to the level of their GDP per inhabitant. Such is the case of Berkshire in advanced services (in both computer and R&D services) and Greater London in operational and traditional services.

The Austrian regions again depict negative deviations, showing less development in every category of business services than expected according to their GDP per inhabitant, whereas the Finish region of Uusimaa contrasts with Åland. Ile de France is placed above the regression line in operational and traditional services, but not in advanced services owing to the weak presence of R&D in that region (as outlined in the second section).

Finally, Table 6.4 reports the correlations between GDP per capita and business services for each country and each category of business services under two scenarios: including all the regions at NUTS2 level (a) and excluding the capital region (b). The main conclusion is related to the substantial dependence of the correlation on the inclusion/exclusion of the capital region. In fact, when the capital region variable is dropped from the analysis, a striking decrease in the Pearson coefficient is observed. This result is particularly evident in two countries, France and Finland, where dropping the capital regions (Ile de France and Uusimaa, respectively) results in the appearance of negligible correlations, especially in computer, advanced and traditional services. In the UK and Austria, the role of the capitals is much less prominent in the relationship between business services and the degree of regional income.

*Table 6.4    Correlations between business services and GDP per inhabitant*

|         | Total | | Advanced | | Traditional | | Operational | | Computer | | R&D | |
|---------|------|------|------|------|------|-------|------|-------|------|-------|------|-------|
|         | A    | B    | A    | B    | A    | B     | A    | B     | A    | B     | A    | B     |
| UK      | 0.79 | 0.64 | 0.78 | 0.65 | 0.78 | 0.62  | 0.51 | 0.16  | 0.71 | 0.7   | 0.59 | 0.67  |
| France  | 0.8  | 0.02 | 0.84 | 0.05 | 0.83 | −0.02 | 0.62 | 0.04  | 0.83 | −0.07 | 0.41 | 0.14  |
| Finland | 0.65 | 0.01 | 0.76 | 0.08 | 0.67 | 0.07  | 0.54 | −0.07 | 0.74 | 0.1   | 0.64 | 0.04  |
| Austria | 0.96 | 0.88 | 0.96 | 0.88 | 0.96 | 0.85  | 0.94 | 0.86  | 0.88 | 0.57  | 0.16 | −0.2  |
| Belgium | 0.87 | 0.58 | 0.84 | 0.66 | 0.85 | 0.51  | 0.83 | 0.34  | 0.66 | 0.12  | 0.1  | −0.37 |

*Note:*   A correlations including all regions at the NUTS 2 level,
         B correlations excluding capital regions.

## FINAL REMARKS

The empirical analysis carried out here seems to reconcile two a priori contradictory results, namely that business services trajectories in European regions are strongly shaped by the very national conditions (the so-called 'national effect') on the one hand, and the importance attached to internationalization (the 'international' effect on the other). In relation to this, whereas the aggregate figures seem to favour the 'national effect' over the 'international effect', both effects may be relevant when looking at different categories of business services.

Furthermore, the density variable and the number of patents per 1000 inhabitants turn out to be outstanding locational factors both when national effects and international effects are controlled for. The analysis confirms that, generally speaking, the business service economy is closely linked to urban density, national profiles and other regulatory and instructional factors which cannot be measured, but, at the same time, to the degree of exposure to international competition.

The correlation between business services and GDP per head is positively biased by the presence of big international cities, the ones most exposed to international competition. Once they are dropped from the analysis, the relationship becomes very weak. Indeed, capital regions benefit from the highest presence of business services as they are the most global regions, even though national characteristics contribute decisively to a further explanation of the differences amongst these capital regions. In this sense, the UK is characterized by a strong presence of business services, and this conclusion is found in many UK regions, particularly in the South East, the leader in advanced business services. This result is due not only to the role of London, but also to the impressive share of business services in the surrounding regions. In contrast, the relative low performance of the Austrian regions can be mainly explained by the different mode of business service production, which is based on a strong production of in-house business services. France is clearly characterized by the role of Ile de France and the relative laggard positions of some regions, such as Centre–Est. Belgium is also characterized by the strong weight of the capital, Brussels. Belgian regions are quite advanced services-oriented, especially in terms of other countries' regions.

To conclude, it must be stated that the analysis has some limitations. First of all, data availability explains why the comparative results are made up of just five countries: France, UK, Austria, Belgium and Finland. In this sense, it would be interesting to have available results from other countries. Thus, as a first step, this information should be obtained for all EU member states and, as a second step, the data collection should be completed with other variables. As a result, more statistics on business services would prove a better analytical framework and would be more useful for any agent interested in them: business service providers, professional associations and policy makers at regional, national or EU level. In the second place, services categories are rather aggregated in some cases (for example, some advanced innovative services cannot be isolated) and a further desegregation would be very valuable in achieving more accurate conclusions. Finally, the explanatory factors included in this chapter are just an example of the likely factors affecting the development of business services, and other variables could have been included in the analysis.

## NOTES

1.  This chapter is a result of the project on 'Services and Innovation' financed by the Ministry of Science and Technology (Ref.: SEC2000-0806). The authors thank Eurostat for the direct provision of unpublished data.
2.  In fact, advanced business services comprise computer services (General Industrial Classification of Economic Activities of the European Community (NACE) 72), R&D (NACE 73), quality control (NACE 74·3) and personnel services (NACE 74·5). On the other hand, traditional services include the many different activities under NACE 74·1 (legal services, auditing, accountancy, and management consultancy), engineering services (NACE 74·2), advertising (NACE 74·4) and other business services (NACE 74·8). Operational services include real estate activities (NACE 70), renting activities (NACE 71), security services (NACE 74·6) and cleaning services (NACE 74·7).
3.  Business services promote 'multinational' firms but not real 'global' firms, which deal more with very standardized manufacturing goods. The few worldwide global firms follow different patterns in their globalization processes.
4.  These places are likely to be relatively close to a central urban area, so that clients and providers can meet physically by making short trips; it takes place in a region with no central place inside.
5   Since the United Kingdom is the most business-oriented services economy, it is not surprising that the dummies representing national effects report negative coefficients, as shown.
6   To be exact, it should be noted that R&D as considered here is only provided by specialized firms and, accordingly, the total R&D expenditure (not covered here) may follow a much more concentrated pattern.

## REFERENCES

Aharoni, Y. (2000), 'The role of reputation in global professional business services', in Y. Aharoni and L. Nachum (eds), *Globalization of Services: Some implications for theory and practice*, London and New York: Routledge.

Bonamy, J. and A. Valeyre (1994), 'Services, relation de service et organisation', in J. Bonamy and N. May (eds), *Services et mutations urbaines: questionnements et perspectives*, Paris: Anthopes.

Daniels, P.W. (1993), *Service industries in the world economy*, Oxford: Blackwell.

Daniels, P.W. and F. Moulaert (eds) (1991), *The Changing Geography of Advanced Producer Services*, London: Belhaven Press.

Daniels, P.W., S. Illeris, J. Bonamy and J. Phillippe (eds) (1993), *The Geography of Services*, London: Frank Cass.

De Bandt, J. and J. Gadrey (eds) (1994), *Relations de service, marchés de services*, Paris: CNRS Editions.

Dunning, J.H. (1993), *Multinational Enterprises and the Global Economy*, Harrow: Addison-Wesley.

European Commission (2001), *Barriers to Trade in Business Services*, Centre for Strategy and Evaluation Services, European Commission Report.

European Economy (2000), 'Mergers and acquisitions', *European Economy*, Supplement A, Economic Trends, no. 5/6.

Gago-Saldaña, D. (2001), 'Information technologies and their effects on internal and external economies: the case of European services.', unpublished summer dissertation, University of Warwick.

Gross, R. (2000), 'Knowledge creation and transfer in global service firms', in Y. Aharoni and L. Nachum (eds), *Globalization of Services: Some implications for theory and practice*, London and New York: Routledge.

Illeris, S. (1994), 'La localisation des producteurs et utilisateurs de services', in J. Bonamy and May, N. (eds), *Services et mutations urbaines: questionnements et perspectives*, Paris: Anthopos.

Illeris, S. (1996), *The Service Economy: A Geographical Approach*, Chichester, UK: John Wiley & Sons.

Marshall, N. and P. Wood (1995), *Services and Space: Key Aspects of Urban and Regional Development*, Singapore: Longman Singapore Publishers.

OECD (2001), 'Information and Communication Technologies and Rural Development', OECD, Paris.

Petit, P. (1986), *Slow growth and the service economy*, London: Francis Printer.

Riddle, D.I. (1986), *Service-led Growth: The role of the service sector in world development*, New York: Praeger.

Roberts, J. (1998), *Multinational Business Service Firms*, Aldershot: Ashgate.

Rubalcaba-Bermejo, L. (1999), *Business Services in European Industry – Growth, Employment and Competitiveness*, Luxembourg: European Commission, DGIII-Industry.

Rubalcaba-Bermejo, L. and J.R. Cuadrado-Roura (2002), 'Internationalization of Service Industries: A Comparative Approach', in J.R. Cuadrado-Roura, L.J. Rubalcaba-Bermejo and J.R. Bryson (eds), *Trading Services in the Global Economy*, Cheltenham, UK and Northampton, MA, USA: Edward Elgar.

Wood P. (ed) (2001), *Consultancy and innovation: The business service revolution in Europe*, London and New York: Routledge.

PART IV

Internationalization and Innovation: the Challenge for Countries and Regions

# 7. From market to resource-oriented overseas expansion: re-examining a study of the internationalization of UK business service firms

**Joanne Roberts**

## INTRODUCTION

This chapter explores the continuing development of international activity in the business services sector. Although the providers of business services are becoming increasingly international in scope, the process of internationalization within the sector is far from fully appreciated. To explore the developing pattern of internationalization within the sector the main findings of a study of the internationalization of UK business service firms conducted in the early 1990s are considered. Evidence suggests that business service firms expand through various stages into overseas markets to supply existing clients and to seek out new clients. More recently, however, technological advances are facilitating the growth of resource-oriented internationalization in a number of service sectors. The purpose here is to reconsider the finding of this earlier research in the light of recent technological developments and to assess the prospects for the rise of resource-oriented internationalization in the business services sector.

Business services are used ultimately by business firms and other productive enterprises. They are extremely diverse, including activities concerned with handling tangible products, such as machinery repair or catering, and providing intangible expertise, like accountancy or consultancy services. The focus of this research is on knowledge-intensive business services (KIBS), specifically advertising, accountancy, public relations, market research, computer services and management consultancy services. These subsectors display high levels of internationalization, and potential for further overseas expansion. Furthermore, they embody a number of common characteristics which act as constraints on the methods of internationalization utilized. These include, for example, the need for personal contact between producer and client, the importance of quality and reputation, a long-term buyer/seller relationship, human capital and knowledge intensiveness, and the need for cultural sensitivity.

The chapter begins with a brief review of the internationalization of business services before an outline of the main findings arising from a study of the internationalization of UK business service firms is provided. The growth of resource-oriented international service activity is then considered. This is followed by an assessment of the prospects for the rise of resource-oriented internationalization in the business services sector. Finally, conclusions are drawn and suggestions for further research are outlined.

## INTERNATIONALIZATION OF BUSINESS SERVICES

The internationalization of business services has attracted much attention in recent years (O'Farrell and Wood, 1998; Bagchi-Sen and Sen, 1997; O'Farrell *et al.*, 1995, 1996; Aharoni, 1993; Aharoni and Nachum, 2000). Although the overseas expansion of business service firms dates from the late nineteenth century and early 1900s,[1] this process clearly gained momentum during the 1970s and 1980s. The development of large multinational advertising, accountancy, computer service and consultancy firms, such as WPP Group Plc, PricewaterhouseCoopers, EDS and McKinsey Co., respectively, provide evidence of the increasing levels of international activity in the sector.

According to UNCTC (1990) the international expansion of business service firms is the result of both demand- and supply-driven forces. With the increasing globalization of economic activity, business service firms have come under growing pressure to follow their multinational clients. The influence of multinational clients on the internationalization of service firms has been widely recognized and empirically confirmed (Weinstein, 1974; Dunning, 1989; Esperanca, 1992).

In addition, the general restructuring of production within the manufacturing sector, with firms concentrating their efforts on core activities while buying in peripheral intermediate goods and services rather than producing them in-house, is leading to a rising demand for producer services. This process of externalization is of particular significance to the growth of business services firms (Perry, 1992), which can achieve economies of scale and scope when providing services to a large market (Enderwick, 1992). However, a market of sufficient size may only be achieved through internationalization. In this respect, the international expansion of business services can be seen as supply-driven, in the sense that firms have sought to reach out to a wider market.

Worldwide exports of business service are increasing (WTO, 2000). European Union export credits arising from 'other business services'[2] increased from a value of US$30 281 million in 1987 to US$129 535 million in 1996, whilst for the United States, over the same period, credits arising from 'other business services' rose from US$10 635 million to US$32 360 million (OECD/EUROSTAT, 1999, pp.76–7). Figure 7.1 traces the expansion of credits arising from 'other business services' for the G7 countries from 1987 to 1996. While credits rose for all of these countries, only the USA and the United Kingdom experienced a persistent surplus over the period, and France benefited from a surplus from 1991 onwards. The other countries experienced persistent deficits. There are, then, clear differences among the advanced nations in terms of the level of international competitiveness in the supply of business services.[3]

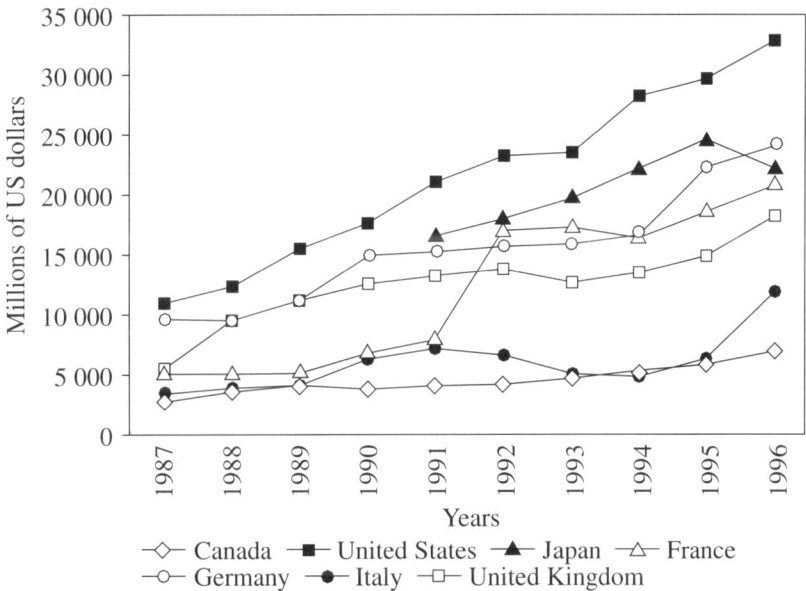

*Source:* Compiled from OECD/EUROSTAT (1999, Table A.10).

*Figure 7.1 Other business service credits by country, 1987–96*

According to data presented by UNCTAD (1999, pp.418–29), foreign direct investment (FDI) in the business services sector is also increasing. For example, the inflow of FDI in the business services sector almost doubled for developed countries, from US$7 262 million in 1988 to US$14 181 million in 1997. For the developing countries, the rise was more dramatic:

from US$7 million to US$5616 million over the same period. Moreover, for the developed countries, business services outward stock of FDI rose from US$18709 million in 1988 to US$165644 million in 1997, rising as a percentage of total stock from 1·8 per cent to 5.2 per cent over the period. Business services FDI inward stock for the developed countries rose from US$13646 million in 1988 to US$171790 million in 1997. For the developing countries, FDI inward stock in business services rose from US$350 million to US$21638 million. Business services FDI accounted for 6·8 per cent of the world total inward stock of FDI in 1997, up from 1·7 per cent in 1988. Furthermore, the business services sector has experienced a significant rise in the annual value of cross-border mergers and acquisitions, from a value of US$3353 million in 1991 to US$39427 million in 1998 (Figure 7.2). Mergers and acquisitions are increasingly important mechanisms facilitating and consolidating internationalization in the sector.

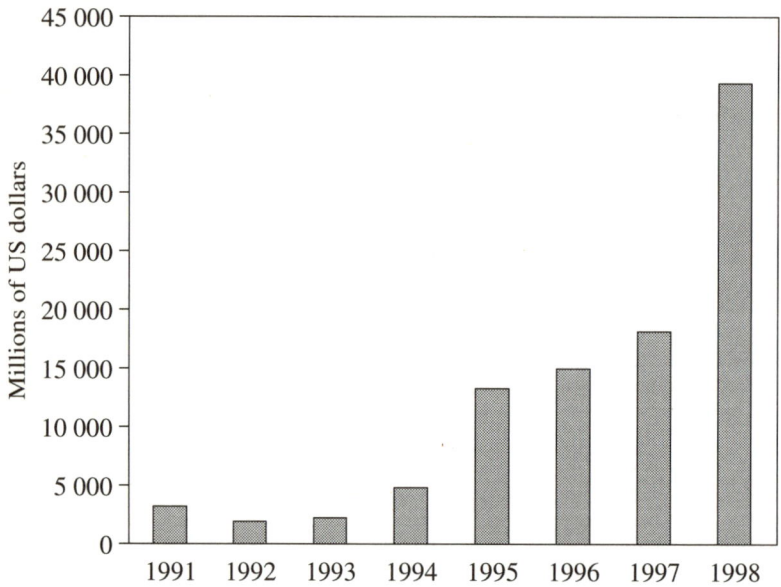

*Source:*  Compiled from UNCTAD (1999, Annex Table B.9).

*Figure 7.2    Value of cross-border mergers and acquisitions in the business service sector 1991–8*

It is important to note that the data reviewed above do not provide a complete reflection of the extent or nature of internationalization. For instance, firms in the sector tend to be skilled labour intensive rather than capital-intensive, hence FDI data do not reveal the true value of overseas facilities in terms of the firm's intangible/knowledge assets. Furthermore, the available data do not reflect international transactions facilitated through the movement of personnel or through intra-firm trade. Additionally, in certain subsectors, international expansion occurs through mechanisms other than FDI or trade; for example, in the accountancy sector, international partnerships are the usual method for the creation of multinational accountancy firms. Importantly, for the purposes of this chapter, FDI data do not reveal the extent to which overseas activity is market- or resource-oriented. For example, to what extent does the rising inflow of FDI in developing countries reflect the search by business service firms for lower-cost labour rather than investment to serve clients in these locations?

Consequently, to develop a deeper appreciation of the internationalization of business service firms, research must include primary as well as secondary sources. For this reason it is useful to review the findings of a research project concerning the internationalization of UK business service firms (Roberts, 1998, 1999).

## THE INTERNATIONALIZATION OF UK BUSINESS SERVICE FIRMS

### A Note on Research Strategy

The research strategy used to study the internationalization of UK business service firms included a variety of methods, from a review of literature and secondary data sources to the collection and analysis of primary data. Primary research involved interviews and correspondence with managers of business service firms and key staff in sector-specific organizations, together with a postal survey of UK business service firms. The postal survey generated a range of useful data. The survey sample was selected from six subsectors: advertising, accountancy, computer services, management consultancy, public relations and market research. A total of 1019 firms were included in the survey conducted between May and September 1990. Usable returns amounted to 328, an overall response rate of 32·2 per cent, providing a sample that reflected the wider population of UK business service firms. Together the various research methods facilitated the compilation of a rich body of primary and secondary evidence through which a useful exploration of the internationalization of firms within the sector could be achieved.[4]

## Key Findings

### Stages in the process of internationalization

In the manufacturing sector the internationalization of the firm often occurs in a number of evolutionary stages (Johanson and Wiedersheim-Paul, 1975; Stopford and Wells, 1972; Aharoni, 1966). For example, on the basis of their research into the internationalization of Swedish firms, Johanson and Wiedersheim-Paul (1975, p.306) argue that internationalization occurs in four distinct stages: (1) no regular export activity, (2) export via independent representatives (agents), (3) sales subsidiary, and (4) production/manufacturing.

Such interpretations of the internationalization process have been criticized (Hood and Young, 1983; Hedlund and Kverneland, 1984). Turnbull (1987), for instance, refutes the evolutionary approach to internationalization, having failed to find supporting evidence in a study of the internationalization of British manufacturing firms. Nevertheless, it is worth considering the extent to which business service firms follow a sequential path in their international development.

Services are widely assumed to be non-tradable owing to characteristics such as intangibility, non-storability and the need for simultaneous production and consumption. They are often referred to as 'locationally bound' (Boddewyn *et al.*, 1986). The assumption that services are non-tradable leads to the view that international expansion occurs in one stage; that is, through the establishment of a presence in an overseas market. However, the study revealed that exports, both independently traded and intra-firm, are integral to the process of internationalization for UK business service firms. Indeed, the business service firms surveyed use a variety of methods of international expansion. Exports and FDI are important, but firms also use joint ventures, franchise agreements and reciprocal arrangements, among other means (Table 7.1).

Consequently, it is possible for business service firms to follow various stages in the process of foreign market entry. As business service firms grow in size and domestic geographical spread, they become international. Internationalization initially takes the form of exports. If the firm provides services in the domestic market for foreign clients, then, clearly, exportation occurs in the form of *domestically located service exports*. Trade frequently arises when overseas customers enter the domestic market, seeking the services available there. For example, the foreign earnings of UK law firms overwhelmingly accrue to the activity of solicitors and barristers in the City of London, where they carry out an essential back-up function to all the main financial service industries (Sowels, 1989). Services may be provided to the overseas client in, for example, a letter or report; here exportation

*Table 7.1   A classification of international activities conducted by business service firms*

Exports
Embodied services, e.g. report, letter, video.
Wired services, e.g. telephone conversation, telecommunications data transfer, Internet-based service.
Domestically located service exports, e.g. legal services provided to a foreign client in the home market.
People-mediated exports, e.g. personnel travelling to overseas market to advise foreign client or to present a report.
Intra-firm exports, e.g. services delivered from the home country to foreign clients via a local office in the overseas market.

Overseas presence*
Export delivery system.
Service production facility, producing services for the local market and perhaps geographically proximate foreign markets.
International production unit – e.g. involved in collecting data on local markets to be used in other countries – management consultants, market research firms; or data entry facility or computer programming unit taking advantage of lower labour costs in overseas market and exporting output back to the home country.
Operations in conjunction with other firms, local or international, to provide services to a particular client, e.g. consortia of firms often used in the computer services sector.

*Notes:*
A business service firm may be involved in any combination of the above within individual markets and between markets.
* These may include wholly or majority owned subsidiaries, joint ventures, franchise or licensing operations, partnerships, associate firms, reciprocal arrangements, together with other methods of local representation.

*Source:*   Roberts (2000a, p.184).

occurs in the traditional sense in the form of *embodied service exports*. Personnel travelling may facilitate *people-mediated exports* and, finally, services can be exported through telecommunication networks in the form of *wired exports*. Evidence from the sample surveyed indicates that UK business service firms use all of these methods of exportation, although people-mediated and domestically located exports prove to be most popular.

From the empirical research, five stages in the internationalization of business service firms can be identified:

1.   provision of services to domestic clients only (no exports),
2.   provision of services to foreign clients in the domestic market (domestically located exports),
3.   provision of services to foreign markets through embodied service exports, people-mediated exports and wired exports,
4.   establishment of a presence through which to deliver a service largely produced in the domestic market (intra-firm exports),
5.   establishment of a service production facility in the overseas market.

Importantly, the number of stages through which a business service firm passes and the length of time spent in each one is variable. Indeed, firms may skip stages, even becoming international in one step through a merger or acquisition. Not only can mergers and acquisitions enable a business service firm to enter a particular market rapidly, but they also facilitate the speedy establishment of international networks. International capabilities can also be acquired rapidly through contractual arrangements. Despite this, the significance of stages, and especially the stage of exporting, in the internationalization of many business service firms must be emphasized.

The research also highlights the presence of intra-firm trade in the sector. Within the survey sample evidence of intra-firm trade was found in relation to the establishment of overseas facilities that function merely to deliver a service largely produced in the domestic market. Moreover, in the wider research the cross-border exchange of intermediate service components was identified. This type of activity suggests the existence, though limited, of international production within the sector. Given the existence of intra-firm trade, there is potential for the development of a more sophisticated form of international production incorporating resource-oriented internationalization. Could it be that a complex system of international production, such as that evident in some manufacturing sectors, in which the location of activities is based on resource cost and availability rather than proximity to the final market, is possible in the business services sector? Or is it possible that the internationalization of business service firms will follow that of the provision of internal service within some multinational firms, in which service production is concentrated at national, regional or international head offices from where services are distributed throughout the organization?

**Factors influencing the pattern of internationalization**
To address some of these questions, a deeper understanding of the internationalization of business service firms is required. Dunning's (1981, 1988) eclectic approach, which draws together three groups of firm-specific advantages arising from ownership, location and internalization (OLI

approach), provides a useful framework with which to explore the existence and development of multinational firms. Ownership-specific advantages include the firm's unique assets such as brands, technology, knowledge base and reputation. Locational advantages relate to the characteristics of over-seas locations including those that are specific to the market, the availability of resources and the general economic and political environment. Finally, internalization advantages arise from securing ownership advantages within the boundaries of the firm, for example, rather than licensing unique knowledge through contractual mechanisms the firm gains greater advantage by internalizing such assets within the firm. Originally developed to analyse the activities of manufacturing firms, the approach was later applied by Dunning (1989) to service firms. Although Dunning's eclectic approach has weaknesses,[5] it does provide a useful framework with which to analyse the international development of business service firms.[6]

*Ownership advantages:* For the firms surveyed, client relations and the quality of staff proved to be the most important sources of competitive advantage, followed closely by use of knowledge and goodwill/reputation (Table 7.2). The firms surveyed did not place great importance on the cost of their services as a source of competitive advantage. The key sources of competitive advantage for the business service firms surveyed are intangible assets. These assets take time to develop, they are difficult to measure and value and, importantly, they are fragile and difficult to protect. Such findings are confirmed by other research exploring the competitive advantages of global business service firms. For example, Aharoni (2000) identifies reputation as the key characteristic influencing global competitive advantage, while Grosse (2000, p.220) notes that competitiveness results from knowledge of clients and relationships with them; global scope of the service firm's affiliate network; methodologies for producing services; knowledge of the market for the services; management skills; and technical/specialized information.

*Locational advantages*: The forces driving the internationalization of business services firms outlined above suggest that the nature of internationalization is largely market-oriented. Indeed, the research findings confirmed the importance of multinational clients in terms of encouraging the international expansion of business services firms. Compared to relations with clients, location is less important as a source of competitive advantage. Exploring the determinants of the location of overseas facilities further underlines the importance of clients as a factor motivating overseas expansion (Table 7.3). It is clear that the overseas expansion of the firms in the sample is primarily market-oriented rather than resource-oriented. For

*Table 7.2    Factors as a source of competitive advantage*

|  | Extremely important (%) | Very important (%) | Important (%) | Of little importance (%) | Not important (%) |
|---|---|---|---|---|---|
| Quality of staff (N = 318) | 78.6 | 13.5 | 2.2 | 0.3 | 5.3 |
| Use of ICT (N = 309) | 19.1 | 46.9 | 31.7 | 12.6 | 8.7 |
| Use of knowledge (N = 319) | 50.8 | 31.3 | 12.2 | 1.6 | 4.1 |
| Location (N = 311) | 7.1 | 15.1 | 42.8 | 23.5 | 11.6 |
| Quality control (N = 311) | 34.7 | 34.1 | 21.9 | 5.1 | 4.2 |
| Client relations (N = 317) | 73.8 | 15.1 | 4.4 | 2.5 | 4.1 |
| Range of services (N = 308) | 14.9 | 30.2 | 38.6 | 11.7 | 4.5 |
| Good will/reputation (N = 318) | 53.8 | 28.0 | 11.9 | 2.8 | 3.5 |
| International coverage (N = 311) | 11.9 | 19.6 | 18.0 | 20.9 | 29.6 |
| Cost (N = 313) | 10.2 | 17.3 | 47.3 | 19.2 | 6.1 |
| Other (N = 14) | 64.3 | 14.3 | 0.0 | 14.3 | 7.1 |

*Source:*    Roberts (1998, p.181).

example, greater importance is attached to client location, market opportunities and potential opportunities than to the availability of skilled labour and the productivity and cost of labour. The evidence suggests that there is little in the way of an international division of labour in the production of the business services studied. From the survey results, then, the scope for resource-oriented overseas expansion in the business services sector would appear to be limited.

*Internalization advantages*:    The research identified a clear preference for overseas expansion through wholly or majority-owned subsidiaries, suggesting that business service firms choose whenever possible to internalize their ownership-specific advantages. The establishment of wholly or majority-owned subsidiaries facilitates control over intangible assets that cannot be fully protected through formal intellectual property rights. The knowledge-intensive nature of the competitive advantages/ownership advantages of business service firms discourages the exchange of such assets in the

*Table 7.3  Factors that determine the location of an overseas presence*

| | Extremely important (%) | Very important (%) | Important (%) | Of little importance (%) | Not important (%) |
|---|---|---|---|---|---|
| Location of client (N = 151) | 43.7 | 24.5 | 16.6 | 5.3 | 9.9 |
| Present market opportunities (N = 148) | 43.2 | 35.1 | 10.8 | 4.1 | 6.8 |
| Potential market opportunities (N = 149) | 41.6 | 32.2 | 16.8 | 4.7 | 4.7 |
| Availability of skilled labour (N = 143) | 17.5 | 19.6 | 30.1 | 15.4 | 17.5 |
| Productivity and cost of labour (N = 138) | 5.8 | 8.0 | 31.9 | 27.5 | 26.8 |
| Knowledge of the country (N = 138) | 19.6 | 26.8 | 29.7 | 12.3 | 11.6 |
| Other (N = 9) | 66.7 | 22.2 | 0.0 | 0.0 | 11.1 |

*Source:*   Roberts (1998, p.195).

market. The transfer of knowledge through market exchange presents a number of difficulties including those explored by researchers studying the economics of information (Arrow, 1969, 1974; Stigler, 1961). Briefly, difficulties arise because of the asymmetric distribution of information concerning the transaction between buyer and seller. The exchange of knowledge gives rise to problems of adverse selection and moral hazard[7] that may prevent such transaction occurring in the open market. Additionally, the ease with which codified knowledge can be reproduced and distributed at low or zero cost tends to undermine private ownership.

**Summary**

From the above analysis it appears that business service firms become international because they possess an ownership advantage with which they can competitively serve clients in overseas markets. Because of the intangible knowledge-intensive nature of their ownership advantages, firms prefer to internalize them within the boundaries of the firm by expanding into overseas markets through the establishment of wholly or majority-owned subsidiaries whenever possible. The location of overseas expansion is determined primarily by the market characteristics: in particular, the location of existing multinational clients. It is also clear that internationalization can occur in a gradual sequential manner, beginning with export and progressing to an overseas production facility over time. However, the environment does have an impact on the costs and benefits of choices regarding the form of internationalization. Local regulation, for example, may prevent the establishment of a wholly-owned foreign firm. Furthermore, technology may influence the nature of services and the form of delivery. The survey upon which the findings reported above are based was conducted in 1990. Since then, there have been major technological developments that may well influence the pattern of internationalization in the sector. In particular, technology is enabling a shift from market- to resource-oriented overseas expansion in some service sectors. Before considering the impact of technological developments on the internationalization of business services, it is useful to explore the growth of resource-oriented expansion in other service activities.

## RESOURCE-ORIENTED INTERNATIONAL SERVICE ACTIVITY

Significant technological developments associated with the convergence of computer and communication technologies have occurred during the last 10 years. These developments have influenced various aspects of the internationalization of services (see, for example, Miozzo and Soete, 2001). The growth of computer-mediated interaction through the Internet and firm-specific Intranets, the rise in use of video conferencing and the adoption of group computer-mediated interaction facilitated through Groupware, together with the falling cost of telecommunications, enable rich multimedia communication across distances. To some extent technology is changing the nature of services. Through embedding services in CD-ROMS they become tangible, they can be stored and the need for simultaneous consumption and production is removed. Similarly, the Internet allows a rich level of interactive communication at a distance, removing, in some instances, the need for close proximity between the producer and client. Such developments have implications for the manner in which service firms supply overseas markets and the scope for international production within service firms.

One consequence of these technological developments is the growth of resource-oriented overseas expansion in sectors such as banking and insurance where, for example, back-office activities and call centres can be located on the basis of cost. In information and computer software services, an international division of labour is well developed whereby firms locate labour-intensive activities in developing countries where suitably skilled labour is relatively cheap compared to the cost in industrialized countries (Lakha, 1994). Customer care and support services that can be delivered by telephone or over the Internet are increasingly being located on the basis of available resources rather than proximity to customers. For example, new opportunities for economic development through the growth of call centres are available in locations as varied as Dublin in Ireland and Bangalore in India (Richardson and Marshall, 1999; Breathnach, 2000; Curtis, 2001). Almost any function that a firm, whether a producer of goods or of services, can outsource from R&D services to payroll and other routine administrative functions, such as transaction processing and customer support services, can potentially be located according to resource availability rather than proximity to the firm's core markets. Such activity may well have contributed to the dramatic increase in the inflow of business service FDI to developing countries noted earlier.

Indeed, developing countries are increasingly promoting themselves as locations for the cost-effective execution of back-office services and other internal services that are open to centralization on a global or regional level. The Philippines government, for instance, keen to attract foreign investment, offers incentives, such as tax holidays, to encourage the inflow of foreign investment in both the manufacturing and service sectors. Service activity conducted by foreign firms in the Philippines includes, for example, AOL, which, with a workforce of 600 Filipinos, processes 10000–12000 e-mail inquiries per day, 80 per cent of which originate in the USA (Wallace and Lucas, 2000, p.6). To illustrate the range of other service activity that can be relocated to developing countries, Table 7.4 lists a number of service activities located in the Philippines.

One might expect these types of services to be relatively well developed in highly competitive markets. Importantly, services that are amenable to such an international division of labour are information-intensive, generally concerned with standardized tasks, the production of which can be assisted with, and delivered through, information and communication technology (ICT). The services provided are centrally concerned with the collection, manipulation and distribution of codified knowledge through standardized processes and procedures. Although produced in a highly prescribed fashion, the services delivered to customers may be highly customized. With the use of ICTs, codified knowledge can be used to produce a vast number of varied outcomes facilitating low-cost customization through the standardization of service components (Quinn and Paquette, 1990; Sundbo, 1994).

*Table 7.4   Examples of service activity located in the Philippines*

| Service activity | Example | Location | Description |
| --- | --- | --- | --- |
| Corporate back office | Caltex | Ortigas | Treasury, human resources |
| Data processing | SPI Technologies | Close to the international airport | Digitize library card catalogues, digitize technical manuals, maintain UK voters' lists, digitize archive for US Dept of Justice |
| Engineering | Fluor Daniel | Alabang | Blueprints, engineering designs |
| Internet service centre | AOL | Clark Special Economic Zone | Answer customer e-mail |
| Medical transcription | Kumar | Makati | Transcribe doctors' dictation |
| Software | Andersen Consulting | Makati | Developing and customizing software |
| Corporate shared services | Procter & Gamble | Makati | Shared accounting functions |
| Corporate technology support | Citibank | Eastwood City | Systems development and support |
| Web site centre | Asian Sources Media | Fort Bonifacio | Maintain large web site |

*Source:*   Adapted from Wallace and Lucas (2000, p.7).

## PROSPECTS FOR RESOURCE-ORIENTED INTERNATIONALIZATION AMONG BUSINESS SERVICE FIRMS

To what extent do the technological advances considered above influence the international development of business service firms? Do they offer opportunities for business service firms to engage in resource-oriented overseas expansion and the development of systems of international production? The prospects for the development of resource-oriented internationalization among business service firms is considered here through an exploration of the process through which business services are produced and delivered.

Resource-oriented overseas expansion among firms is often driven by a desire to reduce the costs of production in order to maintain a competitive market price for the goods or service being produced. While price competition is undoubtedly important in the mass markets for standardized goods and services, the UK business service firms surveyed did not place great importance on the cost of their services as a source of competitive advantage. As noted earlier, client relations and the quality of staff prove to be the most important sources of competitive advantage for the firms surveyed, followed closely by use of knowledge and goodwill/reputation (Table 7.3).

Importantly, clients are often involved in the coproduction of services. Hence the relationship between the provider and the client firm is the key to the successful production and delivery of services. Grosse (2000, p.222) identifies three levels of client relationship that contribute to the service provider's competitive advantage: dealings between individuals at each firm; dealings between the provider's team and the client's team; and long-term, historical dealings that develop client–provider trust and satisfaction in maintaining the relationship. Interpersonal trust is important at all levels and its development requires face-to-face contact. Gadrey and Gallouj (1998, pp.7–8), exploring the interface between provider and customer in business and professional services, also highlight the importance of trust, noting that, unless consultants fulfil their roles correctly, they 'will not succeed in creating an atmosphere of confidence, the key ingredient for keeping a faithful clientele in a profession without tangible products'. Furthermore, analysing the knowledge-intensive business service firm as a distributed knowledge system, Larsen (2001, p.101), highlights the importance to the firm's knowledge base of informal factors such as trust, reputation and personal sympathy between employees and with respect to external collaborators, including client firms.

Technological innovations can help to build trust between the client and producer. In particular, the use of secure Internet communications and investments in the latest supporting technologies can help to build client confidence. Relationships of trust can be developed with the assistance of secure technological systems, certification of standards, membership of professional bodies and so on. Such forms of systems-based and institutional trust are useful to encourage and facilitate the transfer of confidential codified information through electronic facilities (Roberts, 2002). However, in the production and delivery of business services, knowledge transfers are of both a codified and a tacit nature. The transfer of tacit knowledge requires a process of learning that cannot yet be facilitated through electronic information systems. In addition, successful tacit knowledge transfer requires the development of a relationship of trust and mutual understanding between the parties involved (Roberts, 2000b). Face-to-face contact helps to nurture such a relationship between the producer and client, thereby establishing good client relations and optimum service provision.

Face-to-face communication can also eliminate the risks and uncertainties that may accompany communication at a distance whether through computer-mediated communication, telephone or written correspondence. The desire to protect intangible assets such as reputation and the need for absolute confidentiality also encourage colocation between the producer and client. Hence close proximity to the client firm remains crucial in the provision of customized business services, limiting the scope for the development of a system of international production.

Nevertheless, computer service firms are increasingly taking advantage of resource-oriented overseas expansion and international production. The development of software involves the processing of highly codified knowledge, and the back-up services provided to mass software markets are standardized. Hence many computer service firms have located programming facilities and customer support call centres on the basis of cost and availability of suitably skilled labour rather than proximity to client markets. To what extent can these forms of service provision be extended to other business services or, indeed, to the more highly customized activities of computer service firms? In cases where the production and supply of a service require the transfer of both codified and tacit knowledge between the producer and client, international expansion is likely to continue to be primarily market-oriented. The interaction between the client and producer facilitates not only the transfer of knowledge but also its creation (Grosse, 2000; Roberts, 2000c). Gadrey and Gallouj (1998, p.8) view the producer–client interface as a locus and source of innovation: in the sense that, firstly, the creation and evolution of an interface constitutes an innovation in service provision, and, secondly, the interface becomes a laboratory where innovation destined for the client is developed. Also knowledge created in relation to one client can be applied in other projects with that client or with others. In this way business service firms engage in innovative activity and encourage innovation in client firms (Antonelli, 1999; Miles *et al.*, 1995). The colocation of client and producer remains essential for the successful production and delivery of business services.

Consequently, an international system of production focused on minimizing production costs by delivering services at a distance through the use of ICTs seems unlikely to develop in the knowledge intensive business services sector, especially those services dealing with highly specific/idiosyncratic and tacit knowledge. So, although new ICTs provide opportunities for the increased tradability of services, the uptake of these opportunities varies across service sectors. For example, from a study of engineering consultancy, in which much of the knowledge transferred between client and producer is of a codified nature, Baark (1999, p.55) notes that 'the emerging capacity to deliver services in arm's-length transactions across national borders does not appear to be significantly exploited'.

Nevertheless, new technologies are influencing the international development of business service firms. International production can in a sense be seen to be emerging based on the location of skills in the form of centres of excellence. Training centres and location-specific teams of highly mobile personnel enable the knowledge assets embedded in these locations to be shared and exploited internationally. These pools of skills are continuously evolving and may be used in a variety of locations through a combination of computer-mediated communication and movement of key personnel. Centres of excellence are evident in some highly globalized professional business firms. For example, Moore and Birkinshaw (1998) identify a multimedia centre of excellence at Andersen Consulting, which involves a small team of people who have worked on multi-media projects and who know the current technology as well as knowing the firm's experience in the field. They also found a business-to-business marketing centre of excellence that is used to help McKinsey offices develop marketing efforts with clients in many countries. Similarly, Gentle and Howells (1994) note the development of centres of excellence in the European computer services sector. By concentrating expert staff in specific locations, individual and team-based knowledge can be used efficiently throughout the global business service firm. Locationally fixed staff distributed throughout the firm's international network can be supplemented by the temporary mobilization of experts from the firm's centres of excellence. In such a way the scarce resources of highly skilled staff are used efficiently. Local knowledge is used to tailor globally developed solutions to the needs of clients in multiple locations.

In a sense, then, it is possible to see the development of international production based on the location and, indeed, mobility of resources, namely, highly skilled individuals and teams. Though such resources are likely to be located in global cities in advanced countries, they are highly mobile and their skills may be deployed in a wide range of geographical locations both through temporary movement to specific locations and through information networks. Centres of excellence draw on a local knowledge base both within and beyond the boundaries of the firm. Highly skilled personnel absorb local knowledge and facilitate its cross-border distribution through temporary mobilization. Parallels can be drawn between the development of highly creative hubs within the international organizations of business service firms and the concentration of R&D activities in multinational manufacturing firms in a relatively small number of locations. As is the case with international R&D centres, the transfer of knowledge from centres of excellence requires a combination of face-to-face and computer-mediated communication (Howells, 1995; Roberts, 2000b).

In addition to this form of international production based on centres of excellence, the internal functions of business service firms that are not directly concerned with servicing clients but with ensuring the smooth administration of the firm can be located on the basis of cost or availability of resources. Indeed, such internal service activity can be outsourced on a local or global basis.

It may then be possible to see some development of the kind of internationalization that has become more prominent in the computer services sector among other business service sectors. With advances in ICTs, embodied service exports and wired exports are becoming closer substitutes for other forms of service exports and therefore used more widely. Non-client contact exports undoubtedly play a valuable supportive role, enabling the transfer of data, information and codified knowledge. Such exports, to some extent, facilitate other forms of exportation and internationalization through other means. This type of trade is often likely to be intra-firm, rather than direct, and this is an element influencing the international organization of business service firms. Foreign direct investment may take place in certain circumstances to facilitate intra-firm trade. Intra-firm trade supports internationalization through the mobilization of skilled personnel. The challenge for managers of business service firms is to find ways of extending the use of technology at the interface between producer and client without damaging good client relations.

Finally, it is also important to note that there are factors beyond the actual nature of the service that lead to market-oriented rather than resource-oriented overseas expansion. For example, the need for cultural sensitivity and local knowledge is important in the provision of many business services. This leads to a highly fragmented market at an international level. Market-specific customization encourages market-oriented expansion. Moreover, national regulatory requirements have a significant impact on business services (Noyelle and Dutka, 1988). For example, regulation may require that a local presence be capable of production in the local market. Despite movements towards the liberalization of trade in services through the General Agreement on Trade in Services (GATS) and the continuing work of the World Trade Organisation to open up service markets to international competition, many restrictions remain (Findlay and Warren, 2000).

# CONCLUSION

This chapter has reviewed the key findings of a study of the internationalization of UK business service firms, which highlighted the market-oriented nature of the internationalization of business services. These findings have been considered in the light of recent technological developments that have facilitated the growth of resource-oriented internationalization within the service sector. However, resource-oriented service activity is characterized by high levels of standardization. Consequently, such activity is likely to be of limited value when customized and knowledge-intensive business services are being produced. Nevertheless, while core business services activity remains market-oriented, there is scope for resource-oriented activity in the production of highly standardized service components. Moreover, the centre of excellence has been identified as a mechanism used by some business service firms to maximize the effective global deployment of the firm's core knowledge-based assets.

The business service sector is highly polarized, with many small firms serving local markets and a small number of very large firms serving national and international markets. Developments in the patterns of internationalization considered in this chapter suggest that this polarization is likely to continue and, indeed, intensify. While the application of technology in some service sectors, such as retailing, is leading to new entrants and higher levels of competition (for example, as a result of electronic commerce), the international supply of business services is likely to become increasingly concentrated. Barriers to entry are likely to be reinforced by the ability of incumbent firms to use technology in combination with mobile personnel to exploit to the full knowledge-based assets.

Clearly, there is a need for further research into the internationalization of business services and services more generally which addresses the growth of resource-oriented internationalization. To gain an appreciation of such developments it is necessary to develop a research strategy that would involve tracing the international organizational development of business service firms over time. In particular, an examination of the use of electronically mediated communication in both domestic and international interaction with clients would give an indication of the potential for ICT-enabled cross-border business service transactions and the prospects for the development of systems of international production in the sector. Key issues requiring investigation in this area relate to knowledge transfer between client and producers, the role of trust and confidence in client relations, and the establishment and protection of reputation. Increasingly sophisticated forms of electronically mediated communication may become closer substitutes for face-to-face contact. There can be no doubt that, if technological development provides a perfect substitute for face-to-face contact, the pattern of international activity of business service firms, and firms generally, will change radically.

## NOTES

1. Some of today's large accountancy and advertising firms became international at the turn of the century (Leyshon *et al.*, 1987; West, 1987).
2. 'Other business services' include many of the activities that are of concern to this research.
3. The range and quality of data available concerning international transactions in business services are limited. The data that are available must be interpreted with caution since they often lack consistency or comparability across time and between nations.
4. For full details of the research methodology, see Roberts (1998).
5. For a critical discussion of Dunning's eclectic approach, see, for example, Ietto-Gillies, (1992, pp.120–4).
6. For a detailed discussion of the application of Dunning's eclectic approach to the internationalization of service firms, see Roberts (1998). More specifically, West (1996) uses this approach to analyse multinational advertising agencies.
7. Adverse selection is an *ex ante* information problem referring to a situation in which one party in a potential transaction is better informed about a relevant variable in the transaction than the other party. Moral hazard is an *ex post* information problem referring to action which parties in a transaction may take after they have agreed to execute the transaction.

## REFERENCES

Aharoni, Y. (1966), *The Foreign Investment Decision Process*, Boston, MA: Harvard University Press.

Aharoni, Y. (ed.) (1993), *Coalitions and Competition: The Globalization of Professional Business Services*, London: Routledge.

Aharoni, Y. (2000), 'The role of reputation in global professional business services', in Yair Aharoni and Nachum L. (eds), *Globalization of Services: Some Implications for Theory and Practice*, London: Routledge, pp.125–41.

Aharoni, Y. and Nachum L. (eds) (2000), *Globalization of Services: Some Implications for Theory and Practice*, London: Routledge.

Antonelli, C. (1999), *The Microdynamics of Technological Change*, London and New York: Routledge.

Arrow, K.J. (1969), 'Classificatory Notes on the Introduction and Transmission of Technical Knowledge', *American Economic Review*, **59** (2), 29–35.

Arrow, K.J. (1974), *The Limits to Organizatian*, New York: W.W Norton.

Baark, E. (1999), 'Engineering Consultancy: An Assessment of IT-enabled International Delivery of Services', *Technology Analysis and Strategic Management*, **11** (1), 55–74.

Bagchi-Sen, S. and J. Sen (1997), 'The current state of knowledge in international business in producer services', *Environment and Planning A*, **29** (7), 1153–74.

Boddewyn, J.J., M. Baldwin Halbrich and A.C. Perry (1986), 'Service Multinationals: Conceptualization, Measurement and Theory', *Journal of International Business Studies*, **17** (3), 41–57.

Breathnach, P. (2000), 'Globalization, information technology and the emergence of niche transnational cities: the growth of the call centre sector in Dublin', *Geoforum*, **31** (4), 477–85.

Curtis, J. (2001), 'The UK cashes in on Indian support', *Marketing*, 28 June, p.39.

Dunning, J.H. (1981), *International Production and the Multinational Enterprise*, London: Allen and Unwin.

Dunning, J.H. (1988), 'The eclectic paradigm of international production: a restatement and some possible extensions', *Journal of International Business*, **19** (2), 1–31.

Dunning, J.H. (1989), 'Multinational Enterprises and the Growth of Services: Some Conceptual and Theoretical Issues', *The Service Industries Journal*, **9** (1), 5–39.

Enderwick, P. (1992), 'The Scale and Scope of Service Sector Multinationals', in P.J. Buckley and M. Casson (eds), *Multinational Enterprise in the World Economy: Essays in Honour of John Dunning*, Aldershot, UK and Brookfield, US: Edward Elgar, pp.134–52.

Esperanca, J-P. (1992), 'International Strategies in the European Service Sector: A Comparative Study', in M. Casson (ed.), *International Business and Global Integration: Empirical Studies*, London: Macmillan Press, pp.87–120.

Findlay, C. and T. Warren (eds) (2000), *Impediments to Trade in Services: Measurement and policy implications*, London: Routledge.

Gadrey, J. and F. Gallouj (1998), 'The Provider–Customer Interface in Business and Professional Services', *The Service Industries Journal*, **18**, (2), 1–15.

Gentle, C. and J. Howells (1994), 'The Computer Services Industry: Restructuring for a Single Market', *Tijdschrift voor Economische en Sociale Geografie*, **85**, (4), 311–21.

Grosse, R. (2000), 'Knowledge creation and transfer in global service firms', in Y. Aharoni and Nachum L. (eds), *Globalization of Services: Some Implications for Theory and Practice*, London: Routledge, pp.217–32.

Hedlund, G. and A. Kverneland (1984), 'Are Establishment and Growth Patterns for Foreign Markets Changing? The Case of Swedish Investment in Japan', Institute of International Business, Stockholm School of Economics.

Hood, N. and S. Young (1983), *Multinational Investment Strategies in the British Isles: A study of MNEs in the Assisted Areas and in the Republic of Ireland*, London: HMSO.

Howells, J. (1995), 'Going Global: The Use of ICT Networks in Research and Development', *Research Policy*, **24** (2), 169–84.

Ietto-Gillies, G. (1992), *International Production: Trends, Theories, Effects*, Cambridge: Polity Press.

Johanson, J. and F. Wiedersheim-Paul (1975), 'Internationalization of the Firm – Four Swedish Cases', *Journal of Management Studies*, **12** (3), 305–22.

Lakha, S. (1994), 'The new international division of labour and the Indian computer software industry', *Modern Asian Studies*, **28** (2), 381–408.

Larsen, J. N. (2001), 'The Knowledge-Intensive Business Service Firm as a Distributed Knowledge System', *The Service Industries Journal*, **21** (1), 81–102.

Leyshon, A., P. W. Daniels and N. J. Thrift (1987), 'Internationalization of Professional Producer Services: The Case of Large Accountancy Firms', *Working Papers on Producer Services 3*, St David's University College, Lampeter and University of Liverpool.

Miles, I., N. Kastrinos, K. Flanagan, R. Bilderbeek, P. Hertog, W. Huntink, and M. Bouman (1995), *Knowledge-Intensive Business Services: Users, Carriers and*

*Sources of Innovation*, EIMS Publication No. 15, Innovation Programme, Directorate General for Telecommunications, Information Market and Exploitation of Research, Luxembourg: Commission of the European Communities.

Miozzo, M. and L. Soete (2001), 'Internationalization of Services: A Technological Perspective', *Technology Forecasting and Social Change*, **67** (2), 159–85.

Moore, K. and J. Birkinshaw (1998), 'Managing Knowledge in Global Service Firms: Center of Excellence', *Academy of Management Executive*, **12** (4), 81–92.

Noyelle, T.J. and A.B. Dutka (1988), *International Trade in Business Services*, Cambridge, MA: Ballinger.

OECD/EUROSTAT (1999), *Services Statistics on International Transactions 1987–1996*, Paris: OECD/EUROSTAT.

O'Farrell, P.N. and P.A. Wood (1998), 'Internationalization by business service firms: towards a new regionally based conceptual framework', *Environment and Planning A*, **30** (1), 109–28.

O'Farrell, P.N., L. Moffat and P.A. Wood (1995), 'Internationalization by business services: a methodological critique of foreign-market entry-mode choice', *Environment and Planning A*, **27** (5), 683–97.

O'Farrell, P.N., P.A. Wood and J. Zheng (1996), 'Internationalization of Business Services: An Interregional Analysis', *Regional Studies*, **30** (2), 101–18.

Perry, M. (1992), 'Flexible Production, Externalization and the Interpretation of Business Service Growth', *The Service Industries Journal*, **12** (1), 1–16.

Quinn, J.B. and P.C. Paquette (1990), 'Technology in services: creating organizational revolutions', *Sloan Management Review*, **11** (2), 67–78.

Richardson, R. and J.N. Marshall (1999), 'Teleservices, Call Centres and Urban and Regional Development', *The Service Industries Journal*, **19** (1), 96–116.

Roberts, J. (1998), *Multinational Business Service Firms: The Development of Multinational Organizational Structures in the UK Business Services Sector*, Aldershot: Ashgate.

Roberts, J. (1999), 'The Internationalization of Business Service Firms: A Stages Approach', *The Service Industries Journal*, **19** (4), 68–88.

Roberts, J. (2000a), 'The Internationalization of Knowledge Intensive Business Service Firms', in B. Andersen, J. Howells, R. Hull, I. Miles and J. Roberts (eds), *Knowledge and Innovation in the New Services Economy*, Cheltenham, UK and Northampton, MA, USA: Edward Elgar, pp.178–95.

Roberts, J. (2000b), 'From Know-how to Show-how? Questioning the Role of Information and Communication Technologies in Knowledge Transfer', *Technology Analysis and Strategic Management*, **12** (4), 429–43.

Roberts, J. (2000c), 'Knowledge Systems and Global Advertising Services', *Creativity and Innovation Management*, **9** (3), 163–70.

Roberts, J. (2002), 'Trust and Electronic Knowledge Transfer', *International Journal of Electronic Business* (forthcoming).

Sowels, N. (1989), *Britain's Invisible Earnings*, Aldershot: Gower.

Stigler, G.J. (1961), 'The Economics of Information', *Journal of Political Economy*, **69** (3), 213–25.

Stopford, J.M. and L.T. Wells (1972), *Managing the Multinational Enterprise: Organization of the Firm and Ownership of the Subsidiaries*, London: Longman.

Sundbo, J. (1994), 'Modulization of service production and the thesis of convergence between service and manufacturing organizations', *Scandinavian Journal of Management*, **10** (3), 245–66.

Turnbull, P.W. (1987), 'A challenge to the stages theory of the internationalization process', in P.J. Rosson and S.D. Reed (eds), *Managing Export Entry and Expansion*, New York: Praeger.

UNCTAD (1999), *World Investment Report 1999: Foreign Direct Investment and the Challenge of Development*, New York: United Nations

UNCTC (1990), *Transnational Corporations, Services and the Uruguay Round*, New York: United Nations.

Wallace, P. and N. Lucas (2000), 'The Philippines as a Regional Center for Shared Services', *Business Review, the Business Digest of the European Chamber of Commerce of the Philippines*, February (available from *http://www.rent-a-team.com/philippines_as_center.htm*).

Weinstein, A. K. (1974), 'The International Expansion of US Multinational Advertising Agencies', *MSU Business Topics*, Summer, 29–35.

West, D. (1987), 'From T-Square to T-Plan: The London Office of the J.Walter Thompson Advertising Agency 1919–70', *Business History*, **29** (2) 199–217.

West, D.C. (1996), 'The Determinants and Consequences of Multinational Advertising Agencies', *International Journal of Advertising*, **15** (2), 128–39.

WTO (2000), 'Developing countries merchandise exports in 1999 expanded by 8·5 per cent — about twice as fast as the global average', *Press Release*, 6 April (*http://www.wto.org/english/news_e/pres00_e/pr 175_e.htm*).

# 8. Services internationalization: characteristics, potentials and barriers of Danish services firms

**Anders Henten and Torben Vad**

## INTRODUCTION

There is, unfortunately, little statistical material on the internationalization of services. International trade and investment statistics on services are scarce[1] and it was therefore positively received when the Danish Ministry of Industry in autumn 2000 commissioned a study on the internationalization of Danish service industries. The present chapter is based on the results of this study.[2]

The motivation of the Ministry of Industry for commissioning such a study was primarily the new round of WTO (World Trade Organisation) negotiations in relation to which the Ministry sought to gain improved knowledge on the actual potentials and barriers regarding the internationalization of services when involved in international negotiations. Secondly, Denmark witnessed a decrease, larger than the Western European average, in its share of international trade in services during most of the 1990s. However, in 1998–2000, the Danish share had risen again, which was mainly attributable to a sizeable increase in the export of sea transport services.[3] The Ministry also wanted to gain more knowledge concerning the possible reasons for these developments.

For this purpose, questions regarding the characteristics, potentials and barriers with respect to the sale of services nationally and internationally were put to approximately 1600 Danish companies in a questionnaire survey. The present chapter concentrates on the characteristics of services and the perceived barriers to internationalization. It also touches upon the potential for the internationalization of services.

*Table 8.1    Subsectors and number of respondents*

| | |
|---|---|
| Wholesale and retail | 101 |
| Manufacturing | 303 |
| Operational services | 101 |
| Building and construction | 101 |
| Financial services | 200 |
| Communication services | 200 |
| Transport services | 202 |
| Knowledge-intensive services[1] | 203 |
| Other services | 200 |
| Total | 1611 |

*Note:*  [1] Knowledge-intensive services include the following NACE categories: 7210 Hardware consultancy; 7220 Software consultancy and supply; 7230 Date processing; 7240 Database activities; 7260 Other computer-related activities; 7310 Research and experimental development on natural sciences and engineering; 7320 Research and experimental development on social sciences and humanities; 7411 Legal activities; 7412 Accounting, book-keeping and auditing activities/tax consultancy; 7413 Market research and public opinion polling; 7414 Business and management consultancy activities; 7420 Architectural and engineering activities and related technical consultancy; 7430 Technical testing and analysis; 7440 Advertising; 7450 Labour recruitment and provision of personnel; 7484 Other business activities n.e.c.; 8042 Adult and other education not elsewhere classified.

## SURVEY DESIGN

In the design of the survey, respondents were divided into nine subsectors, as shown in Table 8.1, which also lists the number of respondents in each category. The sample of respondents does not entirely reflect the composition of the service sector in Denmark. There is an overrepresentation of subsectors expected to be the most interesting from an internationalization point of view in terms of actual internationalization and potentials. Furthermore, it is likely that firms with international activities would have a greater tendency to answer the questionnaire than firms without international activities because of the subject of the survey.[4]

The number of respondents is considerable and allows for statistical analysis. 4850 firms were randomly selected[5] and there were 1611 answers to the survey, conducted as a combined telephone/Internet survey. Tourism, an important service area, is not included as the focus is on 'outgoing' international sales activities.[6]

Manufacturing firms are included in this survey on services. The reason is that many manufacturing firms produce and deliver not only manufactured goods but also services. The survey has tried to elaborate on this important fact by including manufacturing firms, and asking them about their service activities. It should be noted, however, that it is possible that not all the included manufacturing firms have been able to differentiate between their manufacturing and service activities in their answers. Some of the answers may, therefore, reflect their manufacturing activities to a certain extent.

However, the survey also shows that many firms, which in official statistics are registered as manufacturing firms, regard themselves as service firms (Table 8.2). This applies to 31 per cent of the manufacturing firms in the survey. Conversely, some of the firms in the service areas regard themselves as manufacturing firms, but this mainly applies to wholesale and retail firms – which is understandable as they handle manufactured goods, which may affect the way they regard their business.[7]

*Table 8.2   Services or manufacturing*

|  | Our firm is primarily a service firm (%) | Our firm is primarily a manufacturing firm (%) |
| --- | --- | --- |
| Wholesale and retail | 79 | 21 |
| Manufacturing | 31 | 69 |
| Operational services | 98 | 2 |
| Building and construction | 92 | 8 |
| Financial services | 95 | 6 |
| Communication services | 93 | 8 |
| Transport services | 98 | 2 |
| Knowledge-intensive services | 91 | 9 |
| Other services | 95 | 5 |
| Total | 81 | 19 |

In order to segment the population in the survey further than in subsectors, firms have answered questions regarding firm size, degree of internationalization and mode of internationalization. The survey includes three size categories: small, medium-sized and large firms. Small companies are defined as firms with 1–10 full-time employees.[8] Medium-sized firms have 11–50 full-time employees, and large firms more than 51. In an international context, 'large' firms in this survey are thus not really large. But the size of firms in the different categories reflects the fact that firms in Denmark, in general, are relatively small in size and that service firms generally are even smaller than manufacturing firms.

Regarding the degree of internationalization, the survey differentiates between three categories: home market-oriented firms, defined as firms which do not have any international sales and have not had international sales for the last three years; international firms, which have had international sales for more than five years and which have international sales of more than 10 per cent of their turnover; and a middle group of experimentally international firms, which have had international sales for less than five years or which have international sales amounting to less than 10 per cent of their turnover. Again, the limits have been set relatively low as the degree of internationalization in services is low in comparison to manufacturing.

With respect to modes of internationalization, only three of the traditional four modes of internationalization, dealt with in WTO, for instance, are included: cross-border trade, foreign establishment and temporary presence of natural persons.[9] Movement of customers (as in the case of tourism) is not included as the survey concentrates on 'outgoing' activities. However, firms were also asked whether they are part of a foreign company, in order to include this determinant of their international activities in the survey.

Finally, it should be mentioned that the design of the survey is focused on the potentials and barriers regarding internationalization. Internationalization is, in this sense, seen as an objective in an almost 'mercantilistic' manner. There is, however, no fundamental value judgment that it is necessary for firms to sell on foreign markets. Many service firms are doing fine in their local markets. But seen in the light of a general increase in the internationalization of the economy including services, internationalization – its potentials and barriers – is seen as a challenge to service firms and the public authorities dealing with the policy and regulatory framework for business development.

## CHARACTERISTICS OF SERVICE INTERNATIONALIZATION

Among the respondents to the questionnaire survey, 38 per cent have international sales. When adding the firms that have tried to sell internationally during the past three years but have terminated their international operations, there are 42 per cent that have or have had international sales activities. These figures are lower than for manufacturing industries. This is illustrated by the figures in Table 8.3 showing that 57 per cent of the manufacturing firms in the survey currently have international sales.

As indicated in Table 8.3, firms with international sales mostly rely on export activities, while sales from foreign establishment or temporary presence of natural persons are at a considerably lower level. This runs counter to the general assumption that the internationalization of services, to a very

*Table 8.3   Modes and levels of internationalization*

| | Export | Foreign establish- ment | Temporary presence | No inter- national sales | Respon- dents |
|---|---|---|---|---|---|
| | | | Per cent | | |
| Wholesale and retail | 39 | 10 | 8 | 53 | 101 |
| Manufacturing | 53 | 8 | 6 | 43 | 303 |
| Operational services | 13 | 2 | 4 | 83 | 101 |
| Building and construction | 11 | 1 | 2 | 88 | 101 |
| Financial services | 28 | 22 | 3 | 56 | 200 |
| Communication services | 38 | 11 | 5 | 56 | 200 |
| Transport services | 31 | 7 | 5 | 60 | 202 |
| Knowledge-intensive services | 32 | 8 | 9 | 61 | 203 |
| Other services | 12 | 5 | 4 | 83 | 200 |
| Total | 31 | 9 | 5 | 62 | 1611 |

high degree, depends on non-export activities. It should, however, be borne in mind that some respondents may not differentiate clearly between different kinds of internationalization, especially between exports and temporary presence of natural persons.

The result reflects the fact that most Danish service firms are small and medium-sized enterprises and that smaller firms have a tendency to rely more on exports than on foreign establishment. Export activities are less committing and require less investment. Except for financial services, the internationalization by means of foreign establishments is considerably lower than the export internationalization in all sectors. However, the survey documents that foreign establishment is the mode of internationalization that has experienced the highest growth rate during the past three years.

Despite the – somewhat surprising – result that export activities are so predominant in services, a comparison with manufacturing shows that services are more dependent on foreign establishment. While 53 per cent of manufacturing firms in the survey have export sales and only 8 per cent foreign establishments, the similar figures for services in total are 26 per cent and 9 per cent, when subtracting the manufacturing section from the survey. The rate between export and establishment is thus 6·6 in the case of manufacturing firms and 2·9 in the case of service firms.

Another kind of internationalization activity from the 'outgoing' export, establishment or foreign presence of natural persons, takes place when firms (in the home country – Denmark) are part of foreign firms. Of the firms in the survey with international activities, 10 per cent are part of foreign firms, and this applies to 24 per cent of the communication services firms. This type of international tie has consequences for the 'outgoing' international activities of firms. Among the communication service firms forming part of foreign firms, almost one-quarter make 81–100 per cent of their international sales to other units of the same firm.[10] For firms being part of international conglomerates, there is often a high degree of intra-firm trade.

## POTENTIALS

Potentials for internationalization are reflected in the result that, while 42 per cent of the firms in the survey have or have had international sales during the past three years, there are 65 per cent stating that their services are not specifically designed for the Danish market. This shows that there is a sizable group (23 per cent) would consider that they could internationalize. Furthermore, the survey shows that almost half the firms with international activities do not focus their engagement on specific geographical markets, which indicates a lack of conscious strategy for internationalization, but also entails increasing internationalization potentials if activities become more focused.

Potentials for internationalization are, moreover, reflected in the result that 63 per cent of the firms with international sales expect their foreign activities to increase in the coming three years. There are large differences among the expectations in the different subsectors, with the greatest expectations among communication and knowledge service firms. And although such expectations are predictable, the survey shows that a similar percentage of firms have experienced actual growth in internationalization in the past three years.

The results of the survey also show that as many as 84 per cent of the firms with no international sales state that the Danish market is sufficiently large for their ambitions; even 24 per cent of firms with international activities state this. Together, these results indicate a polarization between firms that are involved in international sales (38 per cent) and a large group of firms that do not have and have not tried (during the last three years) to have international sales (58 per cent). In between, there is a very small group of firms (4 per cent) that have had international activities during the past three years but do not, at present, have any international sales. From a viewpoint seeking to enhance the internationalization of service activities, these figures must be a cause of concern, especially the very low number of firms that have tried to internationalize (without success) and the very high number of firms that do not have any ambitions to internationalize.

This point is underlined by results of the survey, showing that there are very few firms with only recent experience in internationalization. Of the firms with international sales, 70 per cent have more than five years of experience in international markets, and only 4 per cent have less than one year of experience. Although experience must be considered a good thing, the reason for concern is that an excessive amount of new firms focus solely on the national Danish market. While only 4 per cent of firms with international activities in the survey have less than one year of international experience, 7–8 per cent of the total population of service firms in Denmark are newcomers (less than one year of existence).[11]

With respect to major competitors, 64 per cent of firms with international sales consider that their most important competitors are in the Danish market, while 26 per cent say that they are in the Nordic markets and 50 per cent in the European markets.[12] These figures show that even firms with international sales focus primarily on the national market, which has a natural explanation, as approximately 65 per cent of firms with international sales have less than 40 per cent of their sales in international markets.

However, this does not imply that they focus on a sheltered market. The Danish market is also internationalized in the sense both that there is import of foreign services and that foreign firms settle in Denmark. Competition in the national Danish market is, therefore, not necessarily less sharp than in the larger international markets, a point that comes out clearly in another survey showing that firms with international sales do not consider competition to be sharper internationally than in the Danish market.[13]

## FIRM SIZE

The size of firms is clearly correlated with the degree of internationalization and also the mode of internationalization. While 71 per cent of small firms are solely home market oriented, this only applies to 38 per cent of large firms. Furthermore, only 13 per cent of small firms are international, while for large firms there is a similar figure of 39 per cent (see Table 8.4).

Regarding modes of internationalization, small firms rely to a larger extent than larger firms on exports when going international: 85 per cent of the small firms with international sales use the export mode, while the comparable figure is 78 per cent for the large firms. This is not a very significant

*Table 8.4  Size of firms and degree of internationalization*

|  | Only home market (%) | Experimentally international (%) | International (%) | Respondents |
|---|---|---|---|---|
| Small firms | 71 | 17 | 13 | 873 |
| Medium-sized firms | 52 | 24 | 24 | 525 |
| Large firms | 38 | 23 | 39 | 210 |
| Total | 60 | 20 | 20 | 1611 |

*Note:*  The total number of respondents does not exactly add up to 1611 as three firms have not clearly indicated their size.

*Table 8.5  Size of firms and mode of internationalization*

|  | Export (%) | Foreign establishment (%) | Temporary presence (%) | Respondents |
|---|---|---|---|---|
| Small firms | 85 | 13 | 14 | 246 |
| Medium-sized firms | 80 | 22 | 12 | 247 |
| Large firms | 78 | 44 | 16 | 125 |
| Total | 82 | 23 | 14 | 619 |

*Note:*  The total number of respondents does not exactly add up to 619 as one firm has not clearly indicated its size.

difference. The real difference lies in the propensity to internationalize via foreign establishment. Only 13 per cent of the small firms with international sales use this mode of internationalization; the figure for large firms is 44 per cent (see Table 8.5).

None of the above-mentioned results are surprising. Large firms have a higher propensity to internationalize and they establish themselves in foreign countries to a higher degree than small firms. However, when considering the Danish business structure with many small firms, this is part of the explanation for the high number of firms that have no ambitions outside the borders of Denmark. It may also be part of the explanation of why Danish service industries have been losing market shares internationally, except for the large Danish sea transport firms. The consequences of the Danish services industry structure unfold in a situation where services on a global scale become more international. Moreover, this sets the basis

for a realistic discussion of the possibilities for increasing the internationalization of Danish service industries.

## SERVICE CHARACTERISTICS

In discussions on service internationalization, different service characteristics with possible implications for the potentials for internationalization
have been analysed. In this section of the chapter, six such characteristics
are briefly examined.[14]
  One of the most debated characteristics is the attachment of services to
the sale of manufactured goods; that is, services are often sold in connec-
tion with goods. The results of the survey confirm this hypothesis: some 69
per cent of all respondents state that the delivery of their services is closely
related to the delivery of their own or other firms' manufactured goods.
This, of course, applies to the wholesale and retail subsector to a very high
degree, but is an important tendency across the whole range of different services. Furthermore, it applies very strongly to services sold internationally.
Here 81 per cent of firms with international sales state that their services
are closely connected to the sale of manufactured goods, while this only
applies to 64 per cent of the firms with only national sales (see Table 8.6).
  This indicates a greater interdependency between international sales of
services and manufactured goods than is found on the national market, one
reason possibly being that it is more difficult to sell services independently
in foreign markets than in the national market. This is often phrased as 'services following goods' into the international markets. However, it may be
more correctly phrased as there being a strong interrelationship, which also
helps goods into the international markets. The results also emphasize the
often-made assumption that there are a great number of international
service sales hidden in the official statistics on merchandise trade.

*Table 8.6   Degree of internationalization and attachment to manufactured
             goods*

|  | Attached (%) | Not attached (%) | Don't know (%) | Respondents |
|---|---|---|---|---|
| Only home market | 64 | 33 | 2 | 971 |
| Experimentally international | 72 | 25 | 2 | 320 |
| International | 81 | 16 | 2 | 320 |
| Total | 69 | 28 | 2 | 1611 |

The most important traditional characteristic of services – with strong implications for the internationalization of services – is that there is simultaneity in production and consumption and that producers and consumers therefore have to meet face-to-face. With ICTs, this does not necessarily apply to information-intensive services any more. However, for most services it still applies. In the questionnaire survey, a question was included on whether the delivery of services requires physically meeting customers at some point in the process. Unfortunately, the question is not clearly phrased, as it does not say whether a physical meeting has to take place in relation to the delivery of the service or just in general, for example in relation to contract negotiations. Still, the answers to the question are taken as an indication of the above-mentioned characteristic.

The results of the survey confirm this characteristic. A total of 79 per cent of all respondents state that the delivery of services requires meeting physically (see Table 8.7), which with some, but not significant, differences between the different services applies to all service subsectors. The characteristic is also confirmed by the finding that the need for meeting physically is more strongly correlated with the presence of natural persons or foreign establishment than with exports. Moreover, there is a higher percentage of firms with only national sales than firms with international sales that say that meeting physically is necessary, which indicates that services that are less dependent on meeting physically may become internationalized more easily.

*Table 8.7   Degree of internationalization and personal contact*

|  | Necessary (%) | Not necessary (%) | Don't know (%) | Respondents |
|---|---|---|---|---|
| Only home market | 82 | 17 | 0 | 971 |
| Experimentally international | 78 | 22 | 0 | 320 |
| International | 75 | 25 | 0 | 320 |
| Total | 79 | 20 | 0 | 1611 |

A related hypothesis is that the need for local presence in the delivery of services limits the potentials for internationalization. Internationalization by way of foreign establishment or presence of natural persons is an answer to the need for local presence, and the material in the survey confirms the expected correlation between the need for local presence and internationalization via establishment or presence of natural persons.

The material also confirms that there is a negative correlation between the need for local presence and export activities. Export is by far the most important mode of internationalization in the survey and, if examined at a general level including all internationalization modes in total, the home market based firms see a greater need for local presence than the firms with international sales (see Table 8.8).

*Table 8.8    Degree of internationalization and need for local presence*

|  | Need for local presence (%) | No need for local presence (%) | Don't know (%) | Respondents |
|---|---|---|---|---|
| Only home market | 73 | 26 | 1 | 971 |
| Experimentally international | 60 | 39 | 1 | 320 |
| International | 56 | 43 | 2 | 320 |
| Total | 67 | 32 | 1 | 1611 |

A fourth assumption regarding services is that the application of ICTs will lead to greater possibilities for trading information-intensive services over distance and in asynchronous ways, that is without the necessity of meeting physically and without any necessary simultaneousness in production and consumption. The results of the survey confirm this assumption. While 42 per cent of all respondents state that the delivery of services wholly or to some extent take place by way of telecommunication lines (for example, e-commerce), and while the figure for building and construction is down to 22 per cent and for operational services down to 31 per cent, the figure for financial services is 66 per cent, for communication services 57 per cent and for knowledge-intensive services 49 per cent. This shows that to a higher degree than other services information-intensive services use telecommunication lines for delivering their services.

Another angle is that firms with international sales use telecommunication lines for delivering services to a greater extent (54 per cent) than firms confined to the home market (36 per cent) (see Table 8.9). When combining the degree of internationalization with the application of telecommunication lines for the delivery of services, it is apparent that there is a correlation between these two factors. When excluding the sectors that for many years have been more internationalized than the average, that is wholesale and retail, transport and manufacturing, there is an obvious difference in the internationalization of, on the one hand, financial services, communication services and knowledge-intensive services and, on the other hand, operational services, building and construction and other ser-

*Table 8.9   Degree of internationalization and delivery via telecommunication lines*

|  | Usage of telecom lines for delivery (%) | No usage of telecom lines for delivery (%) | Don't know (%) | Respondents |
|---|---|---|---|---|
| Only home market | 36 | 63 | 1 | 971 |
| Experimentally international | 47 | 52 | 0 | 320 |
| International | 54 | 46 | 0 | 320 |
| Total | 42 | 57 | 1 | 1611 |

vices. Furthermore, this difference runs along the same lines with respect to the application of telecommunication lines for the delivery of services. Although this does not constitute any proof of the causality between the delivery of services via electronic means and the process of internationalization, it does indicate that the most ICT-intensive service subsectors are also the subsectors that have witnessed the strongest development in the level of internationalization. Financial services, communication services and knowledge-intensive services would not have experienced their actual level of internationalization without the application of ICTs.

A fifth assumption is that services become more tradable the more standardized they are. Often services are seen as very heterogeneous and individualized, which limits trade in services outside relatively confined areas and environments. However, in the survey, 60 per cent of all respondents answer that the services they deliver are relatively standardized – a significant result, which, however, is supported by other questionnaire survey results.[15]

Financial services are highly standardized according to respondents from the sector (77 per cent) and wholesale and retail likewise (70 per cent), while knowledge-intensive services are considered as much less standardized (38 per cent) and communication services also somewhat less standardized (55 per cent). It may therefore seem that there is no correlation between the degree of standardization and the internationalization of services in the different subsectors, as financial services, wholesale and retail, communication services and knowledge-intensive services all are above average in terms of actual foreign sales. However, when examining the figures on a general level, there is a certain, however insignificant, correlation between the degree of standardization and the degree of internationalization (see Table 8.10)

A final assumption is that the potentials for internationalization are

*Table 8.10    Degree of internationalization and standardization*

|  | Standardized (%) | Not standardized (%) | Don't know (%) | Respondents |
|---|---|---|---|---|
| Only home market | 59 | 40 | 1 | 971 |
| Experimentally international | 56 | 44 | 0 | 320 |
| International | 67 | 34 | 0 | 320 |
| Total | 60 | 39 | 1 | 1611 |

*Table 8.11    Degree of internationalization and niche market orientation*

|  | Niche market orientation (%) | No niche market orientation (%) | Don't know (%) | Respondents |
|---|---|---|---|---|
| Only home market | 37 | 63 | 1 | 971 |
| Experimentally international | 45 | 55 | 0 | 320 |
| International | 55 | 46 | 0 | 320 |
| Total | 41 | 37 | 1 | 1611 |

greater if firms focus on specific niche markets instead of trying to reach all markets. But the correlation between niche market orientation and internationalization cannot be identified on a subsector level in the sense that subsectors that are more internationalized also are more niche market-oriented – except for manufacturing, which is the most international subsector in the survey and also markedly the most niche-oriented.

If examined at a general level, there is a clear correlation between niche market orientation and internationalization (see Table 8.11), but if the manufacturing subsector is subtracted from the survey there is no significant tendency in the material.

## BARRIERS PERCEIVED BY HOME MARKET-ORIENTED FIRMS

The majority of firms in the survey are home market-oriented, without any international sales during the last three years (971 out of 1611 firms). This means that they generally do not have any experience with specific barriers on foreign markets, but they do have opinions about why they have not gone out on the international markets. These opinions are important, as some of the currently home market-oriented firms constitute part of the potential for future internationalization.

The questions addressed to the home market-oriented firms deal with the ambitions and competences of firms to internationalize. The main result is that a vast majority of these firms simply do not have ambitions to internationalize, with 84 per cent of the home market-oriented firms agreeing with the statement that the Danish market is sufficiently large for their firm (see Table 8.12).

*Table 8.12    Reasons of home market-oriented firms not to internationalize*

| | Yes (%) | No (%) | Don't know (%) |
|---|---|---|---|
| The Danish market is sufficiently large for our firm | 84 | 13 | 2 |
| It requires too big changes and investments to achieve international sales | 54 | 39 | 8 |
| The services of the firm are developed and addressed with special emphasis on the Danish market | 50 | 46 | 4 |
| The difficulties in servicing foreign markets are too big | 47 | 41 | 11 |
| The firm does not have the economic resources to market services internationally | 38 | 55 | 7 |
| The competencies of the management and the employees are not sufficient – there are linguistic and cultural barriers | 30 | 63 | 7 |
| We have bad experiences or have heard of bad experiences of other firms with internationalization | 14 | 75 | 10 |

*Note:*    The table is based on 971 answers.

The second important reason for not internationalizing is that it requires too big changes and investments to achieve international sales, and that the services of the firms specifically address the Danish market. It is apparently not bad experiences (own or by hearsay) that hinder firms from going international and neither is it, to any large extent (a self-perceived) lack of competencies. The attitude is mostly due to lack of ambition, and this applies to all subsectors in the survey, although with some differences in degree. In the building and construction subsector, there is an almost unanimous consent that the Danish market is sufficiently large. In communication services, financial services and knowledge-intensive services, about two thirds to three-quarters of the firms agree with the statement, indicating that a number of firms in these subsectors have ambitions to cross the national border.

The size of firms has an influence on the perception of the reasons for sticking to the home market. Small firms agree to a greater extent than large firms with the statements regarding the sufficiently large size of the Danish market, the too heavy requirements for changes and investments, the difficulties of marketing services internationally and the lack of economic resources. However, when it comes to competencies of the management and employees, there are no differences in the answers of small and larger firms. From this, one can conclude that lack of competencies is not a central problem, although 30 per cent does point to a certain problem. One can also conclude that the respondents (firm management representatives) perhaps do not realize that there might be a lack of certain competencies – and both conclusions can be partly right.

But one can also claim that the interpretation of competencies in the questionnaire – linguistic and cultural barriers – is too narrow, and that lack of ambitions to internationalize can be seen as a lack of competencies in the sense that it represents a low awareness of the possibilities to sell on international markets – an awareness that could be upgraded through campaigns and vocational training programmes and so on. Most services have traditionally been confined to local markets, and 'the local spirit' of services continues to exist among many firms.

It should be acknowledged that the reality for many firms is that the national market is actually sufficiently large and that there is no reason to venture out into foreign markets. It is only from an internationalization perspective that this becomes an issue. But if the tendency for services is to become increasingly internationalized and firms operating only in the national markets will meet more international competition, the fact that 84 per cent of home market-oriented firms see the Danish market as sufficiently large for their firm does represent a challenge.

## BARRIERS PERCEIVED BY FIRMS WITH INTERNATIONAL SALES

When examining the barriers for the firms with international sales (international and experimentally international), attention has to be directed at both the possible internal barriers, as in the case of home market-oriented firms, and the external barriers, that is, regulatory and market-oriented barriers in the countries of destination for the services. Many different factors have been examined in the questionnaire survey and are reported in Table 8.13.

*Table 8.13   Factors limiting international sales of service firms*

|  | Yes (%) | No (%) | Don't know (%) |
|---|---|---|---|
| We do not have the economic resources for marketing services internationally | 39 | 56 | 6 |
| Our knowledge of the conditions on foreign markets is insufficient | 37 | 58 | 5 |
| Problems with finding a suitable foreign partner | 28 | 65 | 7 |
| Requirements for local presence limits cross-border trade | 27 | 67 | 6 |
| Regulations on foreign markets are not certain and transparent | 27 | 64 | 9 |
| We do not wish to expand further | 24 | 71 | 5 |
| There are technical barriers to trade in terms of special requirements for certification, authorization, etc | 21 | 73 | 6 |
| Linguistic and international qualifications of management and employees are not sufficient | 19 | 77 | 4 |
| Special sector requirements (e.g. in banking) have limiting effects | 14 | 79 | 8 |
| We experience limits with respect to taxing | 13 | 79 | 8 |
| Local firms receive public economic support | 13 | 77 | 10 |
| Public procurement rules and practices give preference to local suppliers | 13 | 77 | 9 |
| The market is affected by corruption | 10 | 82 | 8 |
| There is import VAT limiting market access | 10 | 82 | 8 |
| There are limitations related to labour market rules and rules for presence of natural persons | 9 | 82 | 8 |
| 'Buy local' clauses have limiting effects | 9 | 82 | 10 |
| Hindrances for ownership (e.g. majority ownership) | 6 | 82 | 11 |
| Marketing rules limit trade | 6 | 88 | 6 |
| Rules concerning the transfer of surpluses limit internationalization | 5 | 82 | 13 |
| Rules concerning limitations in the number of suppliers limit market access | 5 | 87 | 9 |
| There are import quotas limiting market access | 5 | 89 | 5 |
| Hindrances in the execution of preferred mode of operation (e.g. franchising) | 2 | 86 | 12 |

*Note:*   The table is based on 633 answers.

The results of the survey indicate that the firms with international sales perceive the biggest problems to be non-regulatory, mostly internal factors: economic resources, knowledge of foreign markets, problems with finding foreign partners, and lacking the wish to expand. Furthermore, two other points rank high: requirements for local presence (where it is unclear in the survey whether the requirements are regulatory or describe a need for local presence) and lack of transparency and clarity of regulations on foreign markets (which in principle could be due to lack of knowledge). After these, primarily non-regulatory and internal, factors, one finds all the regulatory and market-oriented problems, from technical barriers to trade to hindrances in terms of preferred operation mode.

The overall conclusion must be that the limiting factors are primarily due, not to market conditions or regulatory barriers in the foreign countries, but to insufficiencies in economic resources and knowledge of firms. At the same time, it should be acknowledged that there are actually some regulatory problems to be addressed in international negotiations on trade and investment, and that, when looking at the whole spread of regulatory factors, firms will often face at least one or two of these issues.

In order to get deeper into the factors limiting the current internationalization of Danish service providers, five different variables are examined in the second part of this section: subsectors, differences in national markets, implications of firm sizes, modes of internationalization and degree of internationalization.

With respect to subsectors, there is a relatively large variety in the importance of the different regulatory and market-oriented problems. The building and construction subsector especially is an area with significant problems. This applies in relation to the knowledge concerning foreign markets, the lack of transparency in market regulation, the lack of desire for further expansion, technical barriers to trade, public procurement rules, and 'buy local' clauses. But there are also above-average problems in financial services with regard to taxing and special sector requirements. Furthermore, a large proportion of the wholesale and retail firms state that requirements for local presence limit their cross-border sales of services.

In terms of geographical markets, one would expect to find significant differences between the problems in the different countries and regions, and the material in the survey confirms such an expectation. However, the biggest problems are not only in the countries and regions that would have been expected. The countries and regions where firms face the biggest problems are – not surprisingly – China, Japan, other parts of Asia, Central and Eastern Europe and Africa. However, there are also significant problems in the EU, while Central and South America, the Middle East and Australia and New Zealand are the areas with the lowest incidence of problems – even lower than in the Nordic area.

These results, especially the problems in the EU area, should be noted. However, one of the reasons for the more surprising parts of these results may be that firms simply stay out of faraway countries if problems are anticipated. In the nearby countries, on the other hand, firms have to take up the challenge and enter the markets even if there are problems with regulations and market-oriented hindrances.

Concerning firm size, one would expect small firms to face more problems than larger firms, as their resources for clarifying and dealing with problems will often be more limited than those of the larger firms. Regarding the non-regulatory internal barriers, this is confirmed by the survey – except for the factor that deals with the competencies of the firm management and employees. As in the case with the home market-oriented firms, self-confidence with regard to competencies is as high in small as in large firms. However, with respect to the external factors of both regulatory and market-oriented character, the material in the survey is in direct contradiction to the expectation. For many of the factors listed in the survey, the percentage of large firms facing external problems is twice as high as that for small companies.

One of the reasons for this unexpected result can be – as in the case of different national markets – that smaller firms simply stay out of markets where problems are foreseen, while larger firms have to face these issues. It can also be due to a more systematic registration of problems experienced in larger firms. Finally, a possible reason is that larger firms are more internationalized than small firms and that large firms more often have foreign establishments. With the larger engagement in foreign markets, the higher degree of foreign establishment and the specific problems attached to this mode of internationalization, large firms are bound to face more problems than small firms. There is thus a question of covariance between different variables in the survey.

This issue is confirmed by an examination of the correlation between modes of internationalization and the importance (percentage of firms) of barriers in foreign markets. Firms with foreign establishments and natural persons posted temporarily in foreign countries face regulatory and market-oriented problems more often than firms using only the export mode of internationalization. When examining the internal problems regarding resources and ambitions to internationalize, the result is the opposite; that is, export-oriented firms have the largest problems. However, this again may be explained by the covariance with the size of firms, where small firms are more likely to have problems of an internal character than the large firms.

Finally, there is the issue of degree of internationalization, where the immediate expectation is that firms that are experimentally international would face more serious problems than firms with greater experience in international markets. Again, the survey does not confirm this expectation, the reason probably being that firms with more experience are more likely to have registered and discussed their problems in international markets and, therefore, can report more easily on these problems – exactly as in the case of small and large firms.

## SUMMARY AND CONCLUSION

The survey confirms a number of theoretically based expectations regarding services and their internationalization, but also contains some surprising result, among them the following.

- Almost a third of the firms in the manufacturing category regard themselves as primarily service firms. This illustrates the point that a strict dividing line cannot be drawn between the manufacturing and service sectors. Services are delivered by manufacturing firms and vice versa, which indicates that official statistics on merchandise trade contain a sizable share of services.
- Services are seen as relatively standardized by the firms in the survey, with 60 per cent considering their services to be standardized, which contradicts the theoretical assumption that services are heterogeneous. This result is all the more surprising as 'customization' is a positive buzzword with possible effects on the assessment by respondents of the degree of standardization.
- Export is the predominant mode of internationalization in the survey. One would expect a high degree of internationalization by way of foreign establishment or presence of natural persons, the survey also shows that these two modes of internationalization are more predominant in services than in manufacturing. However, the dominant position of exports in services is surprising.
- External barriers of either regulatory or market-oriented character are not considered to be very important by the firms in the survey with international sales. The internal barriers in the firms themselves are seen as more important, not only by the home market-oriented firms but also by the firms with international sales.

Among the results that confirm general expectations are the following:

- A majority of firms (69 per cent) in the survey state that the delivery of their services is closely connected to the delivery of manufactured goods – and the percentage is even higher for firms with international sales (81 per cent). This indicates a close relation between the sale of goods and services and could be seen as a confirmation of the thesis that international sales of services follow goods. However, it may be more correct to speak of a close relationship than of a primacy of goods.
- A similarly large majority of firms (79 per cent) also confirms that the delivery of their services requires that producers and buyers meet physically. The character of this need for meeting physically is, unfortunately, not clear in the survey, but it emphasizes that many services are dependent on face-to-face contact between producers and consumers.
- A known assumption is that ICTs facilitate trade in information-intensive services. The survey shows that these services are the greatest users of telecommunication facilities and that the information-intensive services are becoming increasingly internationalized. A causal relation between these two results is more than likely.

The two most important results of the survey however, are, that

- a major reason for not internationalizing is simply a lack of ambition to expand internationally among a vast majority of firms with no international sales (84 per cent) and even a considerable proportion of the firms with international sales; and
- the most important limitations on the internationalization of service firms are not external market conditions and regulations in foreign countries but internal resources of firms.

This does not mean, however, that there are no external market-based or regulatory problems. (Furthermore, it is an interesting result that, internally in the EU area, there are still considerable problems to deal with in this respect.) But it means that, if the aim is to achieve a higher rate of service internationalization, it is as much a question of motivating firms to reach out internationally. The main issue is apparently not a lack of competencies in the firms but a lack of ambitions and of economic resources.

There is a polarization between the firms that have international sales and firms without international sales. The middle group, trying to internationalize, is very small. It therefore seems that the present picture is relatively frozen. The reason may be that the small firms do not internationalize to any great extent and that there are many small firms in Denmark. A basic issue is, therefore, the size of firms – even though the survey has documented that small firms do not see bigger problems in the foreign markets than the large firms, and even though other surveys indicate that competition in the national market is considered to be as sharp as on international markets. The reason for staying home is not that the home market is sheltered; the primary reason is that the national or even regional or local market is sufficiently large for the ambitions of many firms.

If the aim is to increase international engagement, the conclusion for policy makers and managers must, consequently, be to focus attention less on the obstacles to market access and equal treatment on national markets and more on the positive potentials for firms in foreign markets. In terms of research activities, this leads to an emphasis on the potentials in foreign markets seen in relation to the resources of firms. Cross-border trade still seems, especially for the smaller firms, to be the most important mode of internationalization. Furthermore, ICTs seem to deliver important tools in the process of internationalization. These are important themes for research on service internationalization.

## NOTES

1.  See, for instance, how little information there is on service trade compared with information on merchandise trade in the *International Trade Statistics 2001* of the World Trade Organization.
2.  This chapter is based on a report written in 2001 by Torben Vad, Anders Henten and Torben Pedersen for the Danish Ministry of Industry, *Internationalisaering af service – potentialer og barrierer* (Internationalization of services – potentials and barriers), Copenhagen: Erhvervsfremme Styrelsen.
3.  Ibid., p.18.
4.  In the collection of answers, the nine different subsectors were filled in with respect to the preset number of respondents in the different subsectors and firm size groups but without any predefined rate between firms with or without international activities.
5.  The selection of firms was taken from Købmandsstanden's firm database, 'CD-Direct', which is a reliable database of Danish firms.
6.  International tourism may be considered as an 'ingoing' international activity, as tourists travel to the exporting countries.
7.  The (small) percentages of firms in the other service subsectors categorizing themselves as primarily manufacturing firms may be partly explained by these firms being part of larger entities working first and foremost in manufacturing. However, pure and simple misunderstanding is also possible.
8.  Single-person firms are not included.
9.  See, for example, WTO Secretariat (1999).
10. These figures should be taken with caution as the total number of firms forming part of foreign firms is low.
11. Danmarks Statistik (1999, p.18).
12. Percentages do not add up to 100 per cent as firms in the survey could tick more than one category.

13. See Anders Henten (1999).
14. Characteristics of services are discussed in, for instance, Anders Henten (1994).
15. See Anders Henten (1999).

# REFERENCES

Danmarks Statistik (1999), *Iværksættere i 1990'erne* (Entrepreneurs in the 1990s), Copenhagen: Danmarks Statistik.
Erhvervsfremme Styrelsen (2001), *Internationalisaering af service – potentialer og barrierer*, Lyngby: Erhvervsfremme Styrelsen.
Henten, Anders (1994), 'Impacts of Information and Communication Technologies on Trade in Services', PhD dissertation, Technical University of Denmark, Lyngby.
Henten, Anders (1999), 'Betydningen af eksportintensitet og konkurrenceudsathed for danske servicevirksomheder' (The importance of export intensity and exposition to competition for Danish service firms), Lyngby: SIC Working Papers, no. 5.
WTO Secretariat (1999), 'An Introduction to GATS', Geneva: WTO Secretariat/Trade in Services Division.
WTO (2001), *International Trade Statistics 2001*, Geneva: WTO.

# 9. Internationalization of knowledge-intensive business services in a small European country: experiences from Finland

**Marja Toivonen**

## INTRODUCTION

In the discussion of the relation between the internationalization of services and the process of innovation, knowledge-intensive business services (KIBS) are of special interest. Several studies have shown that these services play an important role as sources of innovation and as supporters of the innovation activities of other firms (Miles, 1999, pp.92–4 and 2001, pp.13–19; Strambach, 2001, pp.60–5). Some technology-oriented KIBS are also precursors of an international economy based on new information technology (Howells, 2000, p.5).

A good overall picture of internationalization in KIBS branches is, however, still lacking and it can be assumed that country-specific features have a strong impact on the development of internationalization in KIBS. The degree and forms of internationalization in KIBS vary according to national contexts, and these differences are important for understanding the linkages between internationalization and innovation.

The purpose of this chapter is to contribute to the above issue by describing and analysing internationalization of KIBS in Finland – a small country, developing fast in the field of information technology. The chapter is mainly based on the author's own study of KIBS, which was conducted through face-to-face interviews in 2000 (Toivonen, 2001). The study mapped out megatrends in various knowledge-intensive businesses, one theme of the interviews being internationalization issues. The study also aimed at investigating the new qualification requirements demanded by internationalization and other important development trends. In general, this is a point of view that has seldom been adopted in KIBS studies.

The definition of KIBS in the study was based on industrial classifications combined with the statistics of educational levels. The industrial sector of business services was taken as a starting point. Those subsectors in which the level of highly educated professionals was clearly above the average were selected from this basic group. On the basis of this procedure, the following branches were included in the study: computer and related services, research and development (R&D) services, legal services, auditing and accounting services,[1] advertising and marketing services, technical services and management consultancy and personnel recruitment services.[2]

In our study, representatives of 10 professional associations and 87 firms in different KIBS branches were interviewed. Of the professional associations, three represented computer services and two auditing services. Legal services, accounting services, advertising services, management consultancy and technical services were each represented by one association.[3] The distribution of firms interviewed by size and by branch is presented in Table 9.1.[4]

In the selection of the interviewees, the leading firms were weighted to ensure the emergence of development trends that would be relevant for the future.[5] As the sample was not random, the firms studied represented all size categories, which compared to the general size distribution of the Finnish KIBS, means an overrepresentation of larger enterprises (see the following section). However, the material obtained well serves the purposes of this chapter, since the larger enterprises are, almost without exception,

*Table 9.1    Firms interviewed in different KIBS branches, by number of personnel*

| Branch | Number of personnel | | | | | | |
| | 1–9 | 10–49 | 50–99 | 100–249 | 250–499 | 500+ | Total |
|---|---|---|---|---|---|---|---|
| Computer and related | | 6 | 8 | 5 | 5 | 2 | 26 |
| Technical services | 1 | 8 | 2 | 2 | | 2 | 15 |
| **Technological KIBS** | **1** | **14** | **10** | **7** | **5** | **4** | **41** |
| Legal services | 5 | 1 | 2 | 1 | | | 9 |
| Auditing and accounting | 2 | 5 | 2 | 3 | 1 | 1 | 14 |
| Advertising | | 5 | 1 | 3 | 1 | | 10 |
| Management consulting personnel recruitment | 7 | 3 | 1 | 1 | 1 | | 13 |
| **Non-technological KIBS** | **14** | **14** | **6** | **8** | **3** | **1** | **46** |
| **All interviewed firms** | **15** | **28** | **16** | **15** | **8** | **5** | **87** |

already international. In terms of their age, too, the firms interviewed represented a wide variety: from a century-old law firm to Internet consultancy firms established the same year as the interview was made. The majority of the firms were situated in the Helsinki metropolitan area, although a few firms from two other regions were also included in the study.

The study was qualitative in nature, and for this reason no information is available from it concerning the degree of internationalization of the Finnish business service sector in general. Thus, before presenting the findings of this study, a brief description is given in the next section of some other studies that have aimed to map out the quantitative side of internationalization in Finland. The results of our own study are hereafter discussed in the following three sections: first, an overview of the forms of internationalization in KIBS is provided; second, the consequences and obstacles in the internationalization process are examined; and third, the qualification requirements associated with internationalization are explored. The final section draws some conclusions.

## DEGREE OF INTERNATIONALIZATION OF THE FINNISH BUSINESS SERVICE SECTOR

In the KIBS branches defined above, a total of 93790 employees worked in Finland in 1999. This accounts for around 8 per cent of the total labour force in the private sector. As in the other Western countries, KIBS branches in Finland have grown fast during the past few decades. The growth was especially considerable in the latter half of the 1990s: from 1995 to 1999, the number of personnel in these branches increased by 46 per cent. The growth was fastest in computer services, in research and development, in management consultancy and in personnel services. (Statistics Finland, 2001). As has been stated in several studies, KIBS are heavily concentrated in large urban areas (Hermelin, 1997, p.17; Martinelli, 1991, pp.73–4). Of the personnel and turnover of Finnish KIBS companies, around a half are in the Helsinki metropolitan area.

In Finland no quantitative studies of the degree of internationalization particularly in the KIBS sector have yet been conducted. However, there is some information concerning the entire business service sector as well as some KIBS subsectors. Information on the entire business services sector is available in the *SME Barometer* commissioned annually by the Ministry of Trade and Industry. The *SME Barometer* studies the views of small and medium-sized enterprises on the general economic development and the economic performance of these firms by means of telephone interviews. One part consists of internationalization issues, but only related to exports. The target group of the S*ME Barometer* consists of enterprises with no more than 250 employees, and the sample size is well over 4000 enterprises. In 2001, 444 enterprises engaged in the business services sector participated in the study. Around 30 000 enterprises in all operate in the business services, and of these practically all employ fewer than 250 persons; according to the latest data from 1999, only 38 enterprises were larger than this.[6]

According to the 2001 *SME Barometer* (Finnish Ministry of Trade and Industry, 2001), a third of the Finnish business service firms were engaged in exports. The share is smaller than that of the manufacturing firms in the SME sector, of which half conducted exports business. It is, however, larger than that of the small and medium-sized enterprises of all sectors, in which the share of export firms was 23 per cent. This result can be interpreted to support the notion that business services are precursors of internationalization inside the service branches. The largest share of firms having export activities was in computer and related services (50 per cent) and the lowest was in accounting and book-keeping services (3 per cent). The average share of exports of the turnover in those business service firms that were engaged in exports was 23·5 per cent. The other EU countries formed the most important target area, to which 54·9 per cent of exports was directed. The position of exports was, however, altogether minor in business services in relation to the domestic market. When asked to identify the firm's most central market, the distributions of answers in the various business service branches were as shown in Table 9.2.

Although exporting for Finnish business service firms is of little importance – only 5 per cent of the firms regard the international markets as the most important market – still, it is not less important than in small and medium-sized enterprises in general. (However, in this approach too, the significance of exports was greater in manufacturing SMEs.) In relation to the domestic market, business services deviate from the other sectors in that they are more oriented to activities at a national level than small and medium-sized enterprises in general and, correspondingly, the local markets are of minor significance. Within business services, computer services differ from the other services, showing greater shares in both exports and activities at the national level.

More detailed studies have been carried out on exports in the software business, and some of these studies deal also with international cooperation in general. A survey directed at 4452 software firms indicated that, among firms involved with software products, the share of the so-called 'born global' firms is on the increase in Finland as in other countries. Of those firms that were engaged in exports, 43 per cent had started exporting in less than one year from their establishment. The most important

*Table 9.2    Market sector considered the most important, by business
service branch, according to the Finnish SME barometer, 2001
(per cent)*

| Branch | Domestic national | Domestic regional | Domestic local | International | Total |
|---|---|---|---|---|---|
| Computer and related activities | 59 | 16 | 15 | 10 | 100 |
| Accounting and book-keeping | 4 | 29 | 68 | 0 | 100 |
| Auditing services | 17 | 40 | 43 | 0 | 100 |
| Technical services | 44 | 30 | 20 | 6 | 100 |
| Advertising services | 30 | 37 | 30 | 3 | 100 |
| Other business services | 49 | 12 | 36 | 3 | 100 |
| All business services | 38 | 25 | 32 | 5 | 100 |
| All SMEs in manufacturing | 46 | 26 | 18 | 10 | 100 |
| All SMEs | 28 | 26 | 41 | 4 | 100 |

functions in international cooperation were product development and help
desk services (Helsinki University of Technology, 2001).

## FORMS OF INTERNATIONALIZATION IN KIBS

Internationalization has been traditionally examined mainly as an export
activity and through foreign investments. Statistics on international activities and many surveys are still limited to this kind of information. When
applied to service branches, this limitation leads to results showing a low
degree of internationalization. The central characteristics of services –
intangibility and the simultaneity of production and consumption – mean
that the share of direct exports in particular does not reach the level of the
manufacturing sectors. Nowadays there are studies of the service branches
which have also highlighted other forms of international activities, such as
services directed at international client firms in the home country, working
abroad, cooperation and agreement schemes between firms and exploitation of data communication networks in the provision of services (Roberts,
1998, pp.17–20).

The last few years have brought about new features in internationalization, underlining the significance of a broader perspective not only in the
service branches but also in the internationalization of the manufacturing

sector. The increased importance of international capital markets, continuous corporate restructuring, and acquisitions and mergers on an international scale are central features of the so-called 'new economy'. Development of information technology allows, in a novel way, international contacts to be maintained, and it forms an infrastructure for a continuously expanding network-like action. These changes mean that internationalization must be seen all the more, not only as concrete economic actions and forms of trading, but as changes in the business environment and challenges to know-how of firms' labour. These changes also affect firms that do not operate directly on international markets (cf. Castells, 1996, pp.66, 92, 95; Schienstock, 1999, pp. 11–12).

The multiplicity of international activities became evident in our interviews of Finnish KIBS firms. The forms of internationalization described by the interviewees can be grouped as shown below.

1. *KIBS firms' own international economic activities*: exports of products and services, export projects; establishment of foreign subsidiaries; international chaining, acquisitions and mergers.
2. *International economic activities through a client firm*: services for internationally operating domestic client firms; serving foreign clients in domestic markets; international assignments and assignments with an international dimension.
3. *Non equity international activities*: international networks not based on shareholding; foreign expert employees.

Table 9.3 presents a summary of the findings in our study concerning the forms of internationalization. Using the above classification, the significance of different types of international activities in various KIBS branches is evaluated and some more detailed descriptions of international operations are provided. In the last column, ways of internationalizing that are typical of each KIBS branch and the implications of internationalization are characterized in general.

It was indicated above that internationalization in the form of exports is generally lower in service branches than in manufacturing. However, there are subsectors within KIBS in which the share of exports is important and growing. Such subsectors are technology-based KIBS (T-KIBS) in which services contain clearly technical products. In some of these subsectors, product-oriented business plays the key role and the sectors resemble high-technology manufacturing sectors. Software product firms may be mentioned as an example. In our study some firms in this sector gained their main income from exports, and some firms that were distinctly specialized in certain products were international market leaders.

Table 9.3   Forms of internationalization in various KIBS branches in Finland

| KIBS branches | Exports of products and services, export projects | Subsidiaries in foreign countries | International chains, acquisitions and mergers | Services for internationally operating domestic client firms | Services for foreign client firms in the home country | Assignments with an international dimension | International networks not based on shareholding | Foreign experts | General characterization of internationalization and its consequences |
|---|---|---|---|---|---|---|---|---|---|
| Computer and related services (incl. Internet consultancy and content production) | Softwares as export articles, also some service concepts for exports | Software firms have established offices in the most important international IT centres | Some international firms of this sector active in Finland (e.g. IBM, ICL) | Central form of internationalization, Nokia's special position | Foreign clients are one form of internationalization | Internet- and e-business solutions tailored to client firms include international dimensions | Informal networks, 'business fellowship' | Number of foreign software experts on the increase | Share of 'born global' firms is considerable |
| Marketing and advertising (incl. communications consultancy) | Some service products have been developed for international dissemination | Some activity in the adjacent areas (Estonia, Latvia, Lithuania) | International chains came to Finland in the 1980s; the sector is controlled by chains, except retail trade advertising | Clients operating on international markets are important | Major international advertisers use the same advertising chain all over the world | International brands | International cooperation | | International chains play an important role |
| Auditing and book-keeping services | | | 'The Big Five' control half of the auditing market; alliances between 'the big five' and international IT firms exist | Many of the clients of 'the Big Five' firms operate on international markets | Services to subsidiaries of international firms in the home country are important in book-keeping firms | Advisory services for international corporate restructuring are important to auditing firms; in book-keeping, | Medium-sized auditing firms have international 'umbrellas' with a common brand | | International corporate arrangements are important assignments; a polarized sector: 'the Big Five' v. small book-keeping firms |

| Service | Foreign establishments | International ownership / chains | Clients | Other forms of internationalization | International assignments / tasks | Networks | Overall trend |
|---|---|---|---|---|---|---|---|
| Legal services | A few in the adjacent areas (Estonia, Russia) | International chaining within the sector is beginning; the influence of 'the big five' also extends to legal services | Big law firms focused on business law specialize in international clients | Foreign clients are one form of internationalization | international law has an effect on assignments | Networking of small enterprises in the home country in response to fiercer competition; both voluntary and exclusive international networks | International assignments have increased considerably; international competition becoming fiercer |
| Technical services | Export projects of engineering firms have a long history (e.g. Russia, Middle East); Office in countries to which projects have been exported | No international chains; engineering firms of foreign owners are also rare | Client firms engaging in export project, e.g. in construction business | Some foreign planning and consulting tasks | Assignments involving international partners and international law have clearly increased | Networks rare | Sector of conventional export projects, few new forms of internationalization |
| Management consultancy | A few international consulting and recruitment firms; international firms from other branches, e.g. IT business, are also penetrating into consultancy | Even small consultancy firms may play a key role in preparatory stages of exporting; cooperation with public sector organizations supporting exports | | | | Networking of small enterprises in the home country in response to international multisectoral consultancy; also European networks of consultants | Sector is dividing into international multisectoral consultants and small-scale domestic consultants |

In several studies, T-KIBS have been regarded as the most pronounced indicator that the service branches are not laggards in internalization compared to manufacturing (for example, Howells, 2000, p.5). On the other hand, according to our study, some firms in the software sector do not regard themselves as service firms at all, but see themselves as an important part of manufacturing. Although there are already quite a few studies of the convergence of manufacturing and services (Preissl, 2000, p.125; Sundbo and Gallouj, 2000, p.48), economic life is often still perceived in practice as being based on the conventional industrial classification, and manufacturing is seen as having an incomparable position.

To prevent the internationalization of KIBS from being 'explained away' by the kinds of interpretations described above, it is important to go into those forms of exports in more detail in which services play a central role, even though there can be products, too. Among such forms have traditionally been the export projects of engineering firms, which in Finland represent one of the oldest forms of internationalization of KIBS. In purely service-based sectors, which have been called, for example 'managerial KIBS' or 'business-based KIBS' (Werner, 2001, p.50), the newly started commodification of services (Preissl, 2000, p.135; Sundbo and Gallouj, 2000, p.44) may offer opportunities for international distribution. In our study the importance of commodification and modularization became apparent, not only in T-KIBS, but also in legal services, advertising, management consultancy and content production. Commodification of background studies regarding corporate acquisitions, or of advisory services aiming at clarifying a corporate image, may be mentioned as examples of this. However, commodification has so far been mainly limited to domestic use. Predictions about export opportunities were mainly made by content production (non-hardware or non-software materials such as text, voice and pictures for ICT devices) firms. Entertainment concepts and teaching contents were mentioned as examples of possible export articles.

The establishment of offices abroad has generally been considered a more natural form of international business than direct exports in the service branches, because in this way personal interaction between the client firm and the service provider can be guaranteed (Mann and Bargas, 1998). Our study, too, indicated that foreign subsidiaries are common in T-KIBS. Software firms especially had started operations in the most important Western centres of information technology very shortly after their establishment. The largest engineering firms carried out, as a result of their export projects, more permanent activities towards the major export countries on the various continents. In managerial KIBS, there were a small number of foreign offices in sectors with few international chains, particularly in legal services. The offices were usually situated in adjacent areas, among others in Estonia and in Russia.

Activities in several subsectors of business services are more and more concentrated in the hands of international chains (OECD, 1999, pp.46, 113). Chains are particularly important in advertising and in auditing. In Finland, except for advertising in the retail trade, advertising lies almost entirely in the hands of international chains, as there is only one major advertising agency outside the chains. In auditing, the so-called 'Big Five' control nearly half of the Finnish market.[7] On the other hand, in technical services and in legal and book-keeping services, there are no chains at all, and the share of foreign firms and foreign shareholding is also very small in these sectors. In legal services, chaining is about to start, and the interviewees predicted that the development in this sector would follow the earlier trends in auditing.

Mergers and acquisitions have a dual meaning for KIBS in terms of internationalization. In some KIBS sectors, especially in services connected with information technology, acquisitions and mergers are an increasingly common way of becoming international (Roberts, 1998, pp.96–7). Secondly, the need for information, background reports and advice related to major international acquisitions in other sectors is an important factor increasing the demand for KIBS, especially in financial and legal services and in management consultancy. The Finnish KIBS firms interviewed stressed the demanding nature of advisory duties related to international corporate restructuring. Tasks of this kind are very important for the knowledge base of a service firm, because the breadth of experiences gained can be utilized in the future, both in new international assignments and in domestic service tasks. International acquisitions and corporate restructuring are also situations in which the added value produced by KIBS for the client company may be considerable. They can be regarded as examples of a situation in which KIBS produce and carry out important non-technological innovations in global knowledge systems (cf. Howells and Roberts, 2000, p.260).

Client firms operating on international markets, or clients becoming internationalized, have often been seen as the main reason why service firms also must go global (Roberts, 1998, pp.99–100). However, even this demand-based internationalization does not mean that the operations of service firms are passive, 'following the client to the international markets'. In our interviews of Finnish KIBS firms, supporting the clients in internationalization turned out to be a very diversified task. The central position of service firms at preparatory stages related to exports has also been stressed in other studies (OECD, 1999, p.10). According to our study, the activities of consulting firms may comprise feasibility studies of the general applicability of internationalization projects, country-specific surveys on the activity sectors, identification of marketing channels and subcontracting potential, recruitment and training of personnel and technical and operational planning of production. In supporting the internationalization of small enterprises, consulting firms work in cooperation with the organizations enhancing internationalization in the public sector, which have developed commodified services, too, in support of internationalization.

In KIBS sectors that have hardly any exports or where the activities are otherwise oriented to the domestic market, serving foreign clients in the home country is often the activity that connects service firms with international contacts (Roberts, 1998, p.95). The interviews showed that some accounting and book-keeping firms, for example, specialized in serving subsidiaries of international firms domiciled in Finland. Such operations, too, require that a KIBS firm acquire a new kind of know-how and familiarize itself with international law, procedures and culture.

According to our study, concentration on a few key clients provided the alternative for an internationalization strategy based on growth. Some of the firms interviewed considered that rapid growth can easily threaten quality and innovativeness. The quality of work in firms of KIBS type is largely based on a common working culture and learning process, which does not necessarily develop at the same pace as the growth of business. Therefore some of the firms reported that they consciously wanted to stay small, and underlined quality management and key-customer relations. In our study, such firms were found in content production, marketing communications, industrial design and management consultancy. In the words of one interviewee working in the design business: 'to have the world's market leader as a customer is more glorious than growth itself'.

As for international assignments, tasks related to international acquisitions have already been discussed above. Domestic corporate restructuring can also have international dimensions: a foreign financier may be involved and agreements between firms may touch upon foreign parties. One central factor affecting internationalization of the business environments of firms has been Finland's accession to the European Union; this also influences firms operating on the domestic market. A major part of the law linked with corporate activities has now been drafted at an international level; competition law and environmental protection law may be mentioned as examples. All this increases the demand for legal and financial consultancy and requires continuous development of expertise of KIBS operating in these branches. In the same way, in T-KIBS, domestic assignments very often include some kind of international linkage. This can be seen very clearly in computer services and in Internet consultancy while they are building communication networks between firms or systems for e-commerce – functions in which the basic nature is global.

Although non-proprietary networks are not directly part of international business, there is reason to consider them as a form of internationalization of firms, for at least two reasons. First of all, networking of small KIBS firms in the home country is often a reaction to the spread of international chains and thereby to tightening of competition. Secondly, the significance of international networks is on the increase, and intermediate forms have been created between voluntary networks and closer forms of business cooperation. In our study, small management consulting firms and law firms in particular specializing in business law stressed the importance of networks. Small enterprises can complement their range of services through networks; networks can be used for creating new client contacts; and it is possible to widen expertise of one's own, and to adopt best practices (cf. Strambach, 1997, p.28). The forms of networks vary from fully informal cooperation to a common brand, and further to exclusive networks the participant firms of which are liable to use the members of the network as partners in other countries. Networks of the latter kind can already be regarded as a hybrid form, between a network and a chain. Such networks are found, for example, in legal services.

Studies have shown that availability of labour has an effect on firms' decisions concerning their place of establishment in KIBS sectors and in other sectors (Roberts, 1998, p.100). Although there is, according to the interviews, a lack of skilled labour in Finnish KIBS firms too, no foreign units have so far been set up as a result: international offices have been domiciled in central markets and near the centres of expertise of their own field. Instead, firms providing software and other computer services in particular have to some extent recruited foreign experts in Finland to ease the shortage of manpower. Besides alleviating the labour shortage, foreign manpower can also be considered a form of internationalization, since many of the interviewees underlined that it promotes other international activities as well. Interaction between experts from different cultures creates an international atmosphere inside a firm, and the foreign employees can, through their contacts, encourage establishment of a subsidiary of the firm in their home country. Although the international exposure of the personnel of KIBS firms in a small and geographically remote country like Finland is still small-scale, the experiences gained corresponded to those identified by the studies of the importance of multicultural values conducted in international information technology centres (for example, Saxenian, 2000, pp.266–8).

# CONSEQUENCES AND OBSTACLES IN THE INTERNATIONALIZATION PROCESS

The tightening of competition in KIBS branches as a consequence of internationalization has already been referred to above. This, as well as other impacts of internationalization, will be studied in some more detail in this section, as well as the factors that either hinder or slow down the internationalization process. In the new KIBS focused on information technology, particularly in the software product business, the internationalization process itself is novel in nature, or, as a Finnish software entrepreneur put it: 'there is no longer an internationalization *process*'. This statement refers to the fact that orientation towards international markets begins immediately on the establishment of a firm, simultaneously with the development of technology and business activities. Instead of progressing towards internationalization in steps, in the conventional manner, firms are born global (cf. Rönkkö, 2001, pp.83–5). Our study also indicated that new IT firms operating outside the Helsinki metropolitan area could go global directly without attempting first to get a foothold in the domestic centres. These results are in contrast to the perception of a gradual process of internationalization and the role of domestic centres as an intermediate stage in the development, which have both been emphasized in previous studies (see, for example, Roberts, 1998, pp.173, 184, 187).

The tightening of international competition is best seen in the establishment of international chains and large international firms, and in their expansion to an increasing number of KIBS sectors. However, the implications of this phenomenon are not straightforward. In advertising, where the role of chains is well established, the standardization of the global nature of marketing communications was considered a drawback in the interviews. On the other hand, advertising agencies belonging to chains can offer services of even quality to clients all over the world (ibid., p.93). Operating within the same chain is also important in guaranteeing trust in situations in which service firms get very close to the core business of their client firm.

In financial services, in which book-keeping firms and small auditing firms are just becoming professionalized, the dominant position of chains means polarization of the sector. On the other hand, the 'Big Five' play an important role in enhancing know-how in this rapidly developing sector (cf. Howells and Roberts, 2000, p.264; Maula, 1999, pp.305–7). In financial services, large and small KIBS also have rather a clear division of labour, so that 'the Big Five' serve mainly large firms and small internationalizing high-tech firms, while the small-scale auditors and book-keeping firms focus their activities on the domestic SME sector. Furthermore, in polarizing branches a vacuum is easily created in the middle market. Some interviewees had found an opportunity for specialization in this very subsector.

Many services and solutions are local, although the system itself is global. In our study, the need to combine the international and local came up interestingly in the new branches, Internet consultancy and content production for Internet and mobile equipment. At the same time as web solutions and content concepts are developed in these branches, local experts are usually responsible for technical support and acquisition of content. According to the interviews, there is plenty of room for national innovations in these new branches. The national and international dimensions become united in an interesting way in the 'Big Five' too. Cooperation there is contract-based and ownership lies in the hands of domestic actors in each country (Roberts, 1998, pp.125–7). For instance, in most cases a traditional Finnish firm, the operations of which go back for decades, is in the background of the 'Big Five' operating in Finland, and the firms also want to maintain their strong Finnish label.

The big international KIBS are all the more often multisectoral. The 'Big Five' provide not only advice related to financial administration, but also IT solutions and legal services. Big IT firms have expanded their operations into management consultancy. Instead of mere advertising, international advertising chains stress communications services in a broad sense. In addition, they function as Internet consultants and make e-business solutions for their customers. According to the KIBS firms interviewed, the blurring of occupational boundaries is chiefly connected with the fact that the client firms expect a full package of solutions from the service firm. Overall solutions again constitute one factor that underlines the significance of KIBS. Instead of offering a service for a certain point in the client firm's value chain (Moulaert and Daniels, 1991, p.4), the operations of KIBS are connected with the strategy of the firm itself. Depending on which branch the KIBS originate in, they have a standpoint of their own in regard to strategy (Internet strategy, brand strategy, design management and so on), while their aim is a comprehensive survey of the client firm's business and the offer of some overall benefit.

Overall solutions and the emphasis on a strategy strengthen the procedures that have otherwise been demonstrated to be typical of KIBS: close cooperation with the client firm, and cooperation between competitors (Hipp, 2000, p.163; Miles, 1999, pp.94–5). International corporate restructuring was mentioned in interviews as an example of a situation in which 'knowledge is distributed naturally among competitors'. Another example was the training of competitors in a situation in which a solution sold to a client firm is used by several service providers. On the other hand, in big international chains, competition can extend inside a chain. Such coordinated competition was encountered in the advertising sector in Finland.

Although the general trend in KIBS is towards enlarging internationalization, country-specific legislation, norms and culture continue to exercise a strong impact on the development of some KIBS branches (Howells, 2000, p.25; Miles, 2001, p.23). In our study, such sectors turned out to be mainly book-keeping services and some technical services. In book-keeping services the small-scale internationalization that came up in the interviews (besides services provided in Finland for foreign firms) was directed towards the neighbouring countries. Short geographical distances as a factor guaranteeing cultural similarity were also stressed in architectural firms. On the other hand, some Finnish engineering firms have traditionally been engaged in export projects to very distant countries, especially in the areas of construction and the forest industry. As far as engineering firms are concerned, in the current situation their main obstacle to internationalization seemed to be the dramatic contraction caused by the recession at the beginning of the 1990s.

In the future, the internationalization development of engineering and book-keeping firms will be linked in an essential way to the use of information technology. In both of these sectors, electronic data transfer will transform the basic work routines. In book-keeping firms, moving over to paperless book-keeping will bring about electronic archiving that will need huge data-processing capacity. Therefore some kind of receipt service centres are expected to emerge connected to the big banks and global firms providing data-processing services. In engineering, new information technology is in use in project-related inquiries and procurements, but the electronic transfer of design documents to the construction site will take longer, according to the interviews. In international projects, Finnish engineering firms make considerable use of the Internet in training local experts. Electronic learning is an important development task in KIBS, and training via the web is a good example of a situation in which international interaction can clearly be made more efficient by means of technology. However, personal contacts will in general continue to play a significant role in interaction between the client and the service provider.

## QUALIFICATION REQUIREMENTS ASSOCIATED WITH INTERNATIONALIZATION

This chapter has already referred above to the way in which even small-scale internationalization, for example serving foreign client firms in the home country, means a considerable change of action for small KIBS firms. All in all, many qualification requirements connected with internationalization are not necessarily part of the basic professional skills of the personnel of

KIBS firms. The fields of expertise that are central in terms of internationalization can be grouped as follows:

1. country-specific information and knowledge of culture, especially in those geographical areas to which international cooperation is directed;
2. mastery of the routines required by internationalization, for example, information on international legislation and the laws of various countries;
3. know-how related to the international development of one's own sector;
4. combination of business skills with internationalization: strategy, entrepreneurial skills, customer orientation, and marketing;
5. combination of expertise in information technology with internationalization; e-business skills;
6. combination of cooperation and networking skills with internationalization.

Country-specific information, knowledge of culture and mastery of routines related to internationalization are qualification requirements in which the situation of KIBS is very similar to that of internationalizing firms of other sectors. These were the very issues in which KIBS firms interviewed experienced benefit in a concrete way from outside aid provided, for example, by public organizations. Such assistance was considered as important as product development subsidies, and the lack of skilled internationalization consultants was brought up. KIBS assess their own expertise as being deficient especially in international business law, types of agreement and international book-keeping, because these branches have changed very rapidly. Those KIBS having or planning offices abroad called for both country-specific information and support in establishing business structures on site.

KIBS firms' knowledge of international developments in their own sector is generally good. However, the upkeep of professional skills is a challenging task in this respect too, because, according to the KIBS firms interviewed, international development must be monitored on a very large scale, and outside Europe as well. Operating models and innovative ideas may originate from unexpected countries; some specialized activities may be highly advanced also in a country not at the leading edge of the new economy.

The greatest difficulties of KIBS firms in internationalization were related, according to the interviews, to business skills. Deficiencies were detected in the knowledge of business practices, and in practical sales and marketing skills. In regard to this, our KIBS study can be said to reflect to some degree the national characteristics, because the lack of sales and marketing skills has been demonstrated to be a major problem of Finnish enterprises, especially in the SME sector. However, this is partly a more general matter, reflecting the special nature of KIBS: in these branches the establishment of an enterprise is mostly based on a profound interest in the substance itself, rather than on an 'entrepreneurial spirit'. This became evident both in the work of architects, with tens of years of work history, and in the work of software specialists who had just started operations. Combining expertise and entrepreneurship – two very different, partly even opposite, features – is a task in which KIBS must be supported on a practical level, but which should also be studied in more detail. As Tordoir (1995, p.2) puts it: 'The phenomenon of the professionalizing economy is not new . . . What *is* really new and revolutionary, however, is the advent of the professional as an entrepreneur.'

Today international business skills also mean conceiving and utilizing the changes brought about by the new economy. Important issues in this respect are understanding the value chain of a client firm and the changes taking place in it, as well as expertise in electronic business. The change of a client firm's value chain to being more and more customer-based (Hoover *et al.*, 2001, p.37) signifies that KIBS must take their clients' own customers into account as well in their activities. In the KIBS firms interviewed, this aim was attempted by, for example, sector-specific specialization and the active monitoring of changes in consumers' behaviour.

Electronic commerce brings global competition in a new way also into the life of firms operating on the domestic market. In addition, electronic commerce changes the lines of action and earning logics in both KIBS firms themselves and in their client firms. For instance, in advertising, one of the central tasks will be to steer customers to the web sites of firms. The number of free services provided via the Internet will grow, and they will be used as a 'bait' in the creation of more permanent customer contacts. The significance of general familiarity will become increasingly important, which will strengthen the role of international brands.

When examining electronic business in a broader sense, important changes can be detected, for example in various intermediate organizations (Howells, 2000, p.28). Many of the old operators between the producer and the consumer become useless, but at the same time wholly new kinds of intermediaries are emerging. From our study, services provided for mobile equipment arose as an example of activities in which there are separate service producers, service compilers and providers of the services to the customer. Another important question is the use of the intranet/extranet system as an instrument for the integration of production chains, since subcontracting and outsourcing are constantly on the increase. When providing services to big international firms, KIBS must be capable of understanding this technical infrastructure of the networking economy.

Networking and the multisectoral aspect, both inside large KIBS and in cooperation projects of small KIBS, increase considerably the importance of personal and social skills. The challenges are even bigger when cooperation and the crossing of professional boundaries take place at the international level. The KIBS firms interviewed especially mentioned social sensitivity, the ability to encourage others, and the readiness to share knowledge and know-how with others as concrete skill requirements related to the future development described above.

## CONCLUDING REMARKS

Our study shows that the same basic features that have been identified in other studies are also found in the internationalization of Finnish KIBS firms: the degree of internationalization of KIBS is higher than the average in the service sector. Exports and offices abroad are common, especially with T-KIBS. The importance of mergers and acquisitions as forms of internationalization is increasing. Internationalization through client firms plays a key role in all KIBS, but it is particularly important in sectors where the activities are mainly concentrated in the home country. The most significant difference in the situation in Finland compared to the leading Western countries is that there are no large international KIBS originating in Finland; instead, chains of foreign origin hold a strong position in several KIBS branches.

On the basis of the interviews of the Finnish KIBS firms, there is reason to emphasize the versatility of the forms of internationalization and the new ways of acting in international markets. Indirect and non-equity forms of internationalization should be examined in more detail, because there are many new features linked with them. These forms of internationalization – for example, assignments involving an international dimension and international networks – are particularly important in the non-technological KIBS, about which relatively little information is available.

When the direct and indirect forms of internationalization are seen as a continuum and at the same time an intertwining phenomenon, the arguments for the low degree of internationalization of the service sectors become all the more unfounded. The question of local ties is also clearly more complex than is often presented. The results of our study concerning the combination of global and local procedures in enterprises operating at the core of the new economy show that locality can have a very modern content. The international perspective is also, in one way or another, all the more often included in the activities of merely domestic firms.

Internationalization strengthens the strategic meaning of KIBS. Client firms operating on the international markets demand overall solutions from the service firms, not separate services. Respectively, larger KIBS have developed in a multisectoral direction and, in smaller KIBS, networks are an answer to the need for an overall service. The promotion of a division of labour based on positive interaction between large and small KIBS and on mutual benefits is a key issue. In this regard, too, more research information on existing procedures would be needed.

In order to be able to provide overall and strategic solutions, KIBS need to develop further their innovative activities, in the field of both technological and organizational innovations. This is one way in which internationalization influences the development of innovation. However, on the basis of our study, a more detailed and comprehensive analysis of linkages between internationalization and innovation in the Finnish KIBS sector has not been possible. Our study has focused mainly on the forms and general consequences of internationalization in KIBS, and, it is hoped, gives fruitful starting points for further research into innovation linkages.

## NOTES

1. The term 'financial services' is also used in this chapter to refer to auditing, accounting and book-keeping services. Thus it does not in this context mean banking services, which are not included in KIBS in this study.
2. This method of defining KIBS on the basis of statistical classifications is deficient, in that the knowledge-intensive business services inside other branches, for example the advisory services provided by investment banks, are excluded from the study. The definition of KIBS has been dealt with on a larger scale by, for example, Hipp (2000, pp.152–5) and Miles *et al.* (1995, pp.23–31).
3. The number of professional associations varies by branch in the Finnish KIBS sector. The most significant of them were included in the study.
4. In this study research and development were examined as a part of technical services. Therefore there is no separate class for these activities in the table.
5. The new industries were also weighted, because of which computer services are overrepresented in the sample. Internet consultancy and content production related to Internet and mobile terminals were included in the computer services sector.
6. The majority of the big Finnish business service firms are operating in the international markets. The big KIBS firms comprise a few engineering firms fully owned by Finnish shareholders, Finnish and international enterprises in the computer service industry, advertising agencies belonging to international chains and the 'Big Five' engaged in financial consultancy.
7. The following firms, which have their origin in the auditing business but are today firms providing versatile corporate consultancy and for example, legal services, comprise the 'Big Five': Arthur Andersen, Deloitte & Touche, Ernst & Young, KPMG, and PriceWaterhouseCoopers.

# REFERENCES

Castells, M. (1996), *The Rise of Network Society*, Malden, MA and Oxford: Blackwell Publishers.

Finnish Ministry of Trade and Industry (2001), *SME Barometer 1/2001*, Helsinki.

Helsinki University of Technology (2001), 'Software Research in Finland 2001' (*http://www.swbusiness.fi/portal/?cid=650019740*).

Hermelin, B. (1997), 'Professional Business Services Conceptual Framework and a Swedish Case Study', Uppsala Universitet, Geografiska regionstudier no. 30.

Hipp, C. (2000), 'Information Flows and Knowledge Creation in Knowledge-Intensive Business Services: Scheme for a Conceptualization', in J.S. Metcalfe and I. Miles (eds), *Innovation Systems in the Service Economy: Measurement and Case Study Analysis*, Boston, Dordrecht and London: Kluwer Academic Publishers, pp.149–67.

Hoover, W.E. Jr, E. Eloranta., J. Holmström and K. Huttunen (2001), *Managing the Demand–Supply Chain: Value Innovations for Customer Satisfaction*, New York: John Wiley & Sons.

Howells, J. (2000), 'The Nature of Innovation in Services', report presented to the OECD 'Innovation and Productivity in Services', Workshop, 31 October–3 November, Sydney, Australia.

Howells, J. and J. Roberts (2000), 'Global Knowledge Systems in a Service Economy', in B. Andersen, J. Howells, R. Hull, I. Miles and J. Roberts (eds), *Knowledge and Innovation in the New Service Economy*, Cheltenham, UK and Northampton, MA, USA: Edward Elgar, pp.248–66.

Mann, M.A. and S.E. Bargas (1998), 'US International Sales and Purchases of Private Services', *Survey of Current Business*, **78** (10), 71–116

Martinelli, F. (1991), 'Producer services' location and regional development', in P.W. Daniels and F. Moulaert (eds), *The Changing Geography of Advanced Producer Services*, London and New York: Belhaven Press, pp.70–90.

Maula, M. (1999), 'Multinational companies as learning and evolving systems: A multiple-case study of knowledge-intensive service companies. An application of autopoiesis theory', Helsinki School of Economics and Business Administration, Acta Universitatis Oeconomicae Helsingiensis A-154.

Miles, I. (1999), 'Services in National Innovation Systems: from Traditional Services to Knowledge Intensive Business Services', in G. Schienstock and O. Kuusi (eds), *Transformation Towards a Learning Economy*, Helsinki: Sitra 213, pp.57–98.

Miles, I. (2001), 'Taking the Pulse of the Knowledge-Driven Economy: the role of KIBS', in M. Toivonen (ed.), *Growth and Significance of Knowledge Intensive Business Services*, Helsinki: Uusimaa T&E Centre's Publications 3, pp.1–25.

Miles, I., N. Kastrinos, K. Flanagan, R. Bilderbeek, P. den Hertog, W. Huntink and M. Bouman (1995), 'Knowledge-Intensive Business Services: Users, Carriers and Sources of Innovation in European Innovation Monitoring System (EIMS)', EIMS Publication no. 15, Luxembourg.

Moulaert, F. and P.W. Daniels (1991), 'Advanced producer services: beyond the microeconomics of production', in P.W. Daniels and F. Moulaert (eds), *The Changing Geography of Advanced Producer Services*, London and New York: Belhaven Press, pp.1–14.

OECD, (1999), *Business Services: Trends and Issues*, Paris:OECD.

Preissl, B. (2000), 'Service Innovation: What makes it different? Empirical Evidence from Germany', in J.S. Metcalfe and I. Miles (eds), *Innovation Systems in the Service Economy: Measurement and Case Study Analysis*, Boston, Dordrecht and London: Kluwer Academic Publishers, pp.125–48.

Roberts, J. (1998), *Multinational Business Service Firms: The Development of Multinational Organisational Structures in the UK Business Services Sector*, Aldershot: Ashgate.

Rönkkö, P. (2001), 'Growth and Internationalization of Technology-based New Companies: Case study of Eight Finnish Companies', in L. Paija (ed.), *Finnish ICT Cluster in the Digital Economy*, Helsinki: ETLA – The Research Institute of the Finnish Economy, B 176 Series, pp.71–131.

Saxenian, A. (2000), 'Networks of Immigrant Entrepreneurs', in C. Lee, W.F. Miller, M. Gong Hancock and H.S. Rowen (eds), *The Silicon Valley Edge: A Habitat for Innovation and Entrepreneurship*, Stanford, CA: Stanford University Press, pp.248–68.

Schienstock, G. (1999), 'Transformation and Learning: A New Perspective on National Innovation Systems', in G. Schienstock and O. Kuusi (eds), *Transformation Towards a Learning Economy*, Helsinki: Sitra 213, pp.9–56.

Statistics Finland (2001), 'Business Register – Statistics of Enterprises and Establishments', unpublished database, 1990–99.

Strambach, S. (1997), 'Knowledge-intensive services and innovation in Germany', University of Stuttgart, Institute of Geography.

Strambach, S. (2001), 'Innovation Processes and the Role of Knowledge-Intensive Business Services (KIBS)', in K. Koschatzky, M. Kulicke and A. Zenker (eds), *Innovation Networks. Concepts and Challenges in the European Perspective*, Technology, Innovation and Policy 12, Series of the Fraunhofer Institute for Systems and Innovation Research (ISI), Heidelberg: Physica-Verlag, pp.53–68.

Sundbo, J. and Gallouj, F. (2000), 'Innovation as a Loosely Coupled System in Services', in J.S. Metcalfe and I. Miles (eds), *Innovation Systems in the Service Economy: Measurement and Case Study Analysis*, Boston, Dordrecht and London: Kluwer Academic Publishers, pp.43–68.

Toivonen, M.I. (2001), 'Main Development Features of Knowledge Intensive Business Services', in M. Toivonen (ed.), *Growth and Significance of Knowledge Intensive Business Services*, Helsinki: Uusimaa T&E Centre's Publications 3, pp.65–81.

Tordoir, P.P. (1995), *The Professional Knowledge Economy. The Management and Integration of Professional Services in Business Organizations*, Boston, Dordrecht and London: Kluwer Academic Publishers.

Werner, R. (2001), 'Knowledge-Intensive Business Services in the Oulu Region – Business Development and Geographical Linkage', in M. Toivonen (ed.), *Growth and Significance of Knowledge Intensive Business Services*, Helsinki: Uusimaa T&E Centre's Publications 3, pp.49–64.

# 10. Services, scale, and structures of internationalization: northwest England's environmental technologies firms

## Sally Randles and Bruce Tether

## INTRODUCTION

Using data from a survey of environmental technologies and services (ETS) firms in northwest England, this chapter investigates relationships between services, especially knowledge-intensive business services (KIBS),[1] and manufacturing. The analysis also explores the profile of the firms in terms of their patterns and structures of internationalization, with a particular concern for whether the functions firms undertake (such as manufacturing or KIBS or both) relate to the extent and nature of their internationalization. A further dimension of the analysis draws attention to the range and diversity of organizational forms and structures adopted by the subsample of firms engaged in commercial activity across national borders. The chapter considers some implications of these findings for regional 'cluster' development, a cornerstone of current economic development policy in the United Kingdom.

## CLUSTER POLICY IN THE UK

When it came to power in 1997, Britain's Labour government adopted a regional approach to economic development premised on decentralizing decision making and encouraging endogenous economic growth at the subnational 'regional' scale. Eight 'regional development agencies' (RDAs) were established, one for each of the English regions, alongside partial devolution and self-governance via elected assemblies for Scotland, Wales and Northern Ireland. Critics argue that these agencies provide a means by which central government can avoid responsibility for widening regional inequalities in the UK, as responsibility for regional development is passed from central government to the regions. However, as past experience with centrally directed regional policy has at best only reduced the growth of regional inequalities, rather than their extent, an argument can be made in favour of this new approach to regional policy.

The cornerstone of this 'new way of thinking' about regional economic development is the identification of indigenous 'clusters' with 'growth potential' and the potential to stimulate local 'innovation'. 'Clusters' are here defined as 'geographic concentrations of inter-connected firms, specialist suppliers, service providers, firms in related industries and associated institutions' (NWDA, 2000). The development agency for the North West of England (the NWDA),[2] identified seven 'established sectors'[3] and seven 'emerging sectors'[4] as central to the region's current and future competitiveness. This focus on 'clusters' has become prevalent right across the developed world, though there are several very different interpretations of the term and its policy implications.

However, from the specific perspective of the environmental technologies and services sector in England's North West, two sets of fundamental problems associated with 'cluster' policy are addressed in this chapter. First, it uncovers difficulties in identifying appropriate or even realistic boundaries around a coherent, discrete and internally synergetic group of activities which can be used to capture what is in reality a highly diverse group of activities for which the internal synergies and locally based interrelations look questionable. But, second, and of particular importance for this chapter, the findings suggest that, where processes of internationalization come into play, manufacturing and service firms (and, importantly, combinations of both) in ETS play a range of roles in the local 'ETS economy'. Sometimes, indeed, they stimulate endogenous economic growth, but at other times, on the contrary, they contribute to the dissolution of local ties and an effective opening up of the local economy to outside influences, including import penetration, among other factors. Now these points do not, in themselves, constitute a thorough critique of cluster policy. Such policy may indeed have worthwhile objectives such as developing industry–higher education links, technology transfer, mapping and promoting the 'sector' to external audiences and so on. There is, nonetheless, a case to be made for deepening policy makers' understanding of the realities and limits of cluster policy – and of the structures and dynamics which connect the local economy to the international economy in any particular sector. In particular, we need to understand better the role and relative contributions of (1) manufacturing activities, (2) KIBS, and (3) the interplays of manufacturing and KIBS in internationalization and other processes. These issues are discussed through survey analysis, before which we need to consider classification issues as they relate to the ETS 'sector'.

# DEFINING ENVIRONMENTAL TECHNOLOGIES AND SERVICES

The identification by the North West Development Agency of 'environmental technologies and services' as an 'emerging cluster'[5] with 'growth potential' is interesting because this is not an industry or sector as these are conventionally defined. An examination of the UK's current Standard Industrial Classification (SIC) shows that organizations active in these 'environmental technologies and services' are found in many 'industries' which cut across the conventional distinctions, such as between manufacturing and services (See Table 10.1). This should make us question the received definition of industries, and reflect upon the emergence of new industries or sectors. 'Recycling activities' became separately identified for the first time in the 1992 Standard Industrial Classification, but of course there were recycling activities prior to 1992; they were not primarily recognized as such in the statistical framework.

*Table 10.1   Environmental technologies and services in the Standard Industrial Classification*

| Activity | SIC 92 | Conventional classification |
| --- | --- | --- |
| Manufacture of waste disposers | 29.71 | Manufacturing |
| Manufacture of instruments | 33.20 | Manufacturing |
| Manufacture of industrial process control equipment | 33.30 | Manufacturing |
| Recycling of metal waste and scrap | 37.10 | Manufacturing |
| Recycling of non-metal waste and scrap | 37.20 | Manufacturing |
| Collection, purification and distribution of water | 41.00 | Utilities |
| Landfill | 45.11 | Construction |
| Wholesale of waste | 51.57 | Services (Trade Services) |
| Measures relating to the cleanness of water | 74.30 | Services (Business Services) |
| Measuring of pollution | 74.30 | Services (Business Services) |
| Public administration of environment programmes | 75.12 | Services (Public Administration) |
| Waste collection and disposal | 90.00 | Services (Community Services) |
| Activities of environmental movements | 91.33 | Services (Community Services) |

*Note:*   This list is intended to be illustrative rather than exhaustive.

Arguably, the concept of an industrial 'sector' is increasingly problematic. Typically, 'sectors', and especially manufacturing 'sectors', are currently defined in terms of the nature and purpose of their outputs ('the manufacture of X', where X can be, amongst other things, 'soaps and detergents' (SIC92 24.51), 'flat glass' (26.11), 'cutlery' (28.61) or 'motor vehicles' (34.10)). But this is not the inevitable structure of classification, and it is notable that the earliest industries were often classified by their processes rather than by their outputs. For example, in textiles spinning, weaving and finishing were distinguished, rather than the manufacturer of cotton or wool textiles. These 'process-based' distinctions remain for some long-established (but not other) manufacturing industries in the current SIC. Also notable is that, in contrast to manufacturing, services are still essentially defined by their processes, rather than their 'products'. No form of industrial classification will be perfect, but the point here is to emphasize how the widespread adoption (and essentially unquestioning use) of one form of classification, the SIC reflects and reinforces much conventional thinking about industrial development. This is significant because the SIC reifies the distinction between 'manufacturers' and non-manufacturers (for example, mainly services), effectively concealing the 'service activities' of manufacturers. Secondly, the SIC encourages a conceptualization of atomistic agents competing against one another in fixed industries. Interactions between agents and 'industries', other than through market transactions, are obscured.

These issues are brought into focus with the development of new forms of categorization, such as that of 'environmental technologies and services' (ETS). The OECD has provided a 'working definition' of the 'environmental goods and services industry',[6] which covers

> activities which produce goods and services to measure, prevent, limit, minimize or correct environmental damage to water, air and soil as well as problems related to waste, noise and ecosystems. The industry includes both end-of-pipe equipment and cleaner technologies, products and services which reduce environmental risk and minimize pollution and resource use. (OECD, 1999, p.125; see also OECD, 1996a, 1996b)

The OECD (1999) report further recommends that the 'environmental goods and services industry' be analysed at three levels. Level 1 comprises three main groups:

1.  a *pollution management* group (contributing to the limiting, monitoring, reduction or abatement of pollution of air/water/land),
2.  a *cleaner technologies* and products group (equipment, processes and services contributing to the development and implementation of cleaner/more resource-efficient production or processes), and

3.  a *resource management* group (including energy saving and renewable energy technologies).

Level 2 distinguishes (under each category at Level 1) the main categories of environmental protection business activities: production of equipment and specific materials; provision of services, construction, installation and so on. Level 3 comprises the main classes of environmental protection areas: air, wastewater, solid waste, land remediation, noise and vibration abatement and so on. Although it is not clear why levels should be ordered hierarchically as they are, this approach clearly demonstrates a variety of cross-cutting and multiple bases on which the classification of an 'industry' can be made.

It should be clear from the above that the set of activities labelled 'environmental technologies and services' embraces a wide range of technologies, services and 'knowledges'.[7] For example, the 'technoscientific expert knowledges' which contribute to these activities derive largely from biochemistry, electromechanical engineering (primarily for instrumentation) and computation (including both hardware and software). These platform sciences have developed new subdisciplinary specializations in their specific application to environmental problems and issues,[8] and are applied by firms, by publicly owned organizations or by non-commercial agencies, all of which are grouped together primarily because of their application to a wide-ranging set of environment-related problems. These 'technoscientific expert knowledges' are not sufficient, however, particularly in business. For their successful commercial application also requires a coupling with well-developed business knowledge, especially as this relates to an appreciation of the dynamics and operations of markets for the tradable outputs and business applications of the technologies involved. It is often this business knowledge which represents the most significant 'barrier' to commercial developments.

A further difficulty is where to draw the boundaries of the 'industry'; the OECD itself notes the difficulty in exclusively or exhaustively identifying environmental goods and services. This difficulty is partly due to the existing classificatory constructs which mean that, for example, there is no currently agreed methodology that allows the contribution of 'cleaner technologies, products, processes and services' to be captured or measured in a satisfactory way. Consequently, the OECD also recommends supplementing the *principal product/service* method of classification with two additional categories of *secondary* and *ancillary* activities. Here, 'secondary activities' identifies those firms which produce outputs (as goods or services) that are deemed to be actually or potentially useful for environmental protection, but where 'environmental protection' is not the 'principal' purpose of the outputs. Clearly, this raises the problem of how to define 'actually or potentially useful' (are all pumps and valves actually or potentially useful for environmental protection even though most will never be put to this purpose?). Meanwhile, the 'ancillary' category is intended to capture those activities which have environmental protection as their direct or indirect objective, but which are undertaken by a firm to support its main activities. For example, the effluent treatment activities of a chemicals plant, or the energy/water-saving activities of a hotel chain would be ancillary environmental activities. Unless these ancillary activities are 'outsourced' from specialist service providers whose main business is effluent treatment or energy or water saving, the extent of these activities is unlikely to be recognized in the existing statistics, although they may involve considerable resource allocation. There is a clear parallel here with the apparent increase in the economic significance of services. The extent to which this reflects an increased demand for services, or an increased market provision of services (which are now outsourced by firms but which were previously undertaken in-house) is unclear.

The wider point is that, as activities become traded (and therefore tradable and commodified), they tend to become more visible in statistical classifications. The visibility of activities may differ depending upon whether they are undertaken by private firms, or by public and other non-commercial agencies. Notably, the private–public division of expenditure on environmental protection / remediation activities varies widely across Europe. In Germany, for example, the 'public sector' accounts for 95 per cent of expenditure on wastewater treatment, whereas in the UK the wastewater management 'market' is primarily served by private enterprises: the 'public sector' accounts for only 15 per cent of the total expenditure (OECD, 1996c). Clearly, activities that were undertaken by the 'public sector' might be privatized (and vice versa) and a significant proportion of the apparent growth in 'environmental service' activities in the UK may be accounted for by the outsourcing by local authorities to private enterprises of activities previously undertaken by the local authorities themselves. This has often involved the wholesale transfer of local authority departments into the 'private sector', and thus a recategorization of essentially the same organization. With their growing visibility, partly through the increasing trade of goods and services associated with them, but also through the actions of the environmental movements, the aforementioned environmental problems have attracted increasing attention from policy makers (for example, CEC, 1997).

It is apparent that what is labelled 'new' or 'emerging' often involves a recategorization of the old or existing. The subsector of energy management is a case in point. It traditionally includes a range of electrical and mechanical engineering technologies, such as boilers, compressors and refrigeration, but the modification of such artefacts and their incorporation into, for example, Combined Heat and Power (CHP) systems, brings them under the 'new' classification 'environmental technologies and services'. Likewise, biomass as a form of renewable energy builds on technologies of forestry management (classified traditionally as agriculture and forestry) and the burning of wood for fuel (an energy class activity). Many environmental technologies and service activities have a long history, but what can be considered 'new' or recent are the problem-solution combinations which are responding to existing or emerging 'demands'[9] for solutions, drawing together and applying often well-established knowledge or technologies. Of course the knowledge and technologies may also change considerably: windmill technologies are an example of this, being transformed from their ancient origins to current high-tech status. Similarly, instruments, which have a long history, are continually being developed and transformed, most recently with their increasing computerization. But there is more to this emergence than a recategorization of activities. The identification of 'new problems' (in business, domestic and public sector domains) relates to the creation of 'new' markets, and thus new combinations of (existing and new) knowledge. It also provides a new conceptualization for policy making; with policy recognition comes the potential for policy support. Thus the emergence of this classification cannot be removed from its historical–institutional context. But we must also ask what drives and integrates this set of activities, and what are the (variety) of business organizational forms and institutionalized 'business models' associated with it? We now turn to the particular case of 'environmental technologies and services' in the North West of England.

## ENVIROLINK AND THE ENVIROLINK SURVEY

To facilitate the development of the ETS 'sector' in the North West region of England, the NWDA funds Envirolink.[10] This is an 'industry-led' organization, launched in October 2000, charged with developing the region's environmental technology 'sector' into a competitive force to provide sustainable solutions to environmental problems. To achieve this, Envirolink aims (1) to work together (with businesses and other interested organizations) to improve the environment of the North West; (2) to raise the profile of the North West's environmental technologies and services sector in local, national and global markets; (3) to help North West environmental suppliers to find and win new business; (4) to stimulate the formation of partnerships and consortia to address market opportunities; (5) to provide a forum for exchange of knowledge and experience; (6) to improve the competitiveness of the sector; and (7) to assist in links with the North West Development Agency (NWDA) and other regional/national bodies.[11]

Given the lack of existing statistics, the first objective was to gather intelligence on the 'environmental technologies and services sector' and its surrounding institutional context of non-firm organizations, such as universities, specialist business support agencies, and others such as charities with an operating unit in the North West of England. One of the means by which this was undertaken was a postal questionnaire, which was sent to firms and other organizations in the North West region thought to be active in environmental technologies and services.

Taking an inclusive or extensive approach to defining the 'environmental technologies and services',[12] Envirolink compiled a list of candidate organizations from trade directories, exhibitions and contacts. The survey was then undertaken in several rounds, the first of which took place in July 2000. All of the data presented in this chapter are drawn from the response to the first round of the Envirolink survey.[13]

A total of 190 organizations responded to this first round of the Envirolink survey, from a total sample of 600 organizations. Thus a gross response rate of 32 per cent was achieved, which is relatively high for this type of exercise. In all, 24 responses were from universities, registered charities, members' organizations and publicly funded bodies, while 166 were from private businesses or firms. The analysis that follows is confined to firms, as many of the questions were not appropriate for non-businesses, although we recognize that many of the non-firms organizations may be undertaking international activities and may be assisting firms with their international activities.

The firms were asked to declare what proportion of their sales was attributable to 'environmental technologies and services' (ETS).[14] A total of 14 firms declared that none of their current sales were attributable to ETS. Although these firms may have been previously active in ETS, or may have subsequently become active in ETS, they also have been excluded from the analysis that follows. This leaves a sample of 152 firms in the analysis.

Of these 152 firms, only 60 (39 per cent) declared that all of their sales were attributable to ETS, with a further 22 (14 per cent) declaring that more than half (but less than all) of their sales were ETS-related. Meanwhile, 31 (20 per cent) indicated that ETS accounted for less than 10 per cent of sales, and for 24 (16 per cent) ETS accounted for 11–50 per cent of sales. A final group of 15 (10 per cent) were unable to place a figure on the proportion of their sales due to ETS. This raises several points. Firstly, it appears that the extent to which a firm is active in 'environmental technologies and services' is partly a matter of perspective. For example, one firm in the home insulation business attributed less than 5 per cent of its sales to ETS, although it might reasonably be supposed that all its sales were ETS-related.[15] Because the 'sector' is new or poorly defined, it might be difficult to classify activities as being within or beyond its boundaries. Secondly, it suggests that, if a discernible 'cluster' does exist, many firms consider themselves to be only partially rather than wholly within it – most are also active in other 'sectors' or even other 'clusters'. In terms of internal linkages, these might be more salient than the ETS 'cluster' (such as that for chemicals, energy, food and drink or textiles). Although we do not have information on non-respondents, it seems likely that those firms that attributed a large proportion of their sales to ETS, and which were therefore more likely to consider themselves part of an 'ETS cluster', were more likely to respond to the questionnaire.

*Table 10.2   Firm size, by sales and employment*

| By turnover | N (%) | By employment | N (%) |
|---|---|---|---|
| Under £250 000 | 44 (29) | 1–5 | 48 (32) |
| £250 000–£1m | 27 (18) | 6–15 | 27 (18) |
| £1m–£4.9m | 49 (32) | 16–40 | 32 (21) |
| £5–£14.9m | 18 (12) | 41–99 | 25 (16) |
| Over £15m | 14 (9) | 100+ | 20 (13) |

The response to the survey was primarily from small and micro firms (Table 10.2); those with turnovers below £1m and employing no more than 15 people constituted half the sample. But there were also a few large organizations, six having turnovers in excess of £50m and five employing more than 1000 people. This is a fairly typical pattern for any 'industry', such that the direct economic significance of the few larger firms is often at least as great as the combined mass of small and micro firms. But in contrast to standard industry analyses, we should not assume that the large firms are simply scaled-up versions of the small. Nor should we assume that all firms are active in the same markets. Overall, the evidence shows the micro and small firms tend to be more specialized in serving ETS markets than are the larger firms (Table 10.3). Half of the firms with one to five employees attributed all their sales to ETS, whilst over half the large firms with more than 100 employees declared less than half their sales to be due to ETS.

Almost two-thirds of the firms were active in consultancy, with a third engaged in manufacturing,[16] while other widely undertaken 'functions' included contracting (28 per cent) and research and development (33 per cent). If 'knowledge-intensive business services' are defined as being active in at least one of consulting, training, research and development, surveying

*Table 10.3　Firm size and specialization in environmental technologies and services*

| Employment | Some ETS sales, but undefined % | ETS are less than 50% of sales | ETS are 51–99% of sales | All sales are ETS-related | N |
|---|---|---|---|---|---|
| 1–5 | 6 (12%) | 6 (12%) | 9 (19%) | 27 (56%) | 48 |
| 6–15 | 5 (19%) | 7 (26%) | 6 (22%) | 9 (33%) | 27 |
| 16–40 | None | 11 (34%) | 7 (22%) | 14 (44%) | 32 |
| 41–99 | 3 (12%) | 14 (56%) | 4 (16%) | 4 (16%) | 25 |
| 100+ | 1 (5%) | 11 (55%) | 2 (10%) | 6 (30%) | 20 |
| All | 15 (10%) | 49 (32%) | 28 (18%) | 60 (39%) | 152 |

*Note:*　Likelihood ratio chi-square tests = 33.0, 12 degrees of freedom, significance = 0.001.

or monitoring, and laboratory and analytical services, then three-quarters of the firms were KIBS providers. Conventionally, a strong distinction is made between manufacturers and non-manufacturers, but our sample indicates considerable overlap between these activities, with half the manufacturers also undertaking (KIBS) functions (Howells, 2000b).

To analyse the behaviour of the firms, particularly with respect to internationalization, a classification scheme was developed which first divided them between those engaged in manufacturing, those in KIBS, those in both, and those in neither. This last 'residual group' comprised firms that acted as agents, distributors, contractors or operators (of a facility), but which did not manufacture or provide KIBS. From this classification, by far the largest group was of non-manufacturing KIBS providers (88), and we further divided this between those firms that earned all of their income from ETS activities and those that earned some but not all of their income from ETS activities. Thus the following classification was developed:

1. manufacturers not also undertaking KIBS (25);
2. manufacturers also undertaking KIBS (26);
3. non-manufacturers undertaking KIBS, and firms which did not derive all of their income from ETS activities (45);
4. non-manufacturers undertaking KIBS and which attributed all of their sales to ETS activities (43);
5. other – businesses not fitting the above categories (13).

Clearly, this is just one of several possible classifications, and all such classifications have arbitrary elements. Any classification should be for a

*Table 10.4   Firm size and the various firm types*

| Employment | Manufacturing only | Manufacturing & KIBS | KIBS, partly ETS | KIBS, all ETS | N (%) |
|---|---|---|---|---|---|
| 1–5 | 3 (12%) | 9 (34%) | 13 (29%) | 21 (49%) | 48 (32%) |
| 6–15 | 4 (16%) | 5 (19%) | 7 (16%) | 5 (12%) | 27 (18%) |
| 16–40 | 7 (28%) | 5 (19%) | 8 (18%) | 10 (23%) | 32 (21%) |
| 41–99 | 8 (32%) | 5 (19%) | 9 (20%) | 2 (5%) | 25 (16%) |
| 100+ | 3 (12%) | 2 (8%) | 8 (18%) | 5 (12%) | 20 (13%) |
| All | 25 | 26 | 45 | 43 | 152 |

*Note:*   Likelihood ratio chi-square tests = 26.1, 16 degrees of freedom, significance = 0.053.

purpose, and we are particularly concerned with whether the functions the firms undertook (such as manufacturing or KIBS) related to the extent and nature of their internationalization activities. The small size of the data set restricted the number of dimensions of discrimination that could be used if the number of observations in each group were to remain large enough for statistical analyses

Assessed by size, Table 10.4 shows that the non-manufacturing KIBS firms that specialize in ETS activities tend to be very small, with half having no more than five employees, whilst those that manufacture but do not undertake KIBS tend to be larger. All the categorizations, however, include some very small firms and some large firms.

**Internationalization amongst the Environmental Technologies and Services Firms**

We now turn to the questions of the spatial distribution of markets and, in particular, internationalization. Most of the firms in our sample derived at least some of their sales from the rest of the UK; only a small minority were only active in the North West's regional markets. About 60 per cent earned income from 'overseas' activities, with another small proportion planning to enter international markets in the near future. A third of the firms, however, had no international sales and had no plans to enter international markets. These were mainly non-manufacturing KIBS firms, especially those that derived all of their sales from ETS markets. (See Table 10.5)

We can also examine the spread by region of the firms' overseas sales (Table 10.6). Of those with international sales, seven out of 10 were active in Western Europe, which, since the introduction of the Single European Market in 1992, may be considered a large 'domestic market'. However,

*Table 10.5    Location of sales of ETS firms ( per cent )*

| Sales made | Manufacturing only | Manufacturing & KIBS KIBS | KIBS, partly ETS | KIBS, all ETS | All |
|---|---|---|---|---|---|
| Only locally | 8 | 4 | 13 | 16 | 11 |
| To rest of UK | 92 | 96 | 80 | 79 | 85 |
| To overseas markets | 84 | 77 | 58 | 44 | 61 |
| Of non-internationalized: Planning to internationalize | 0 | 8 | 7 | 14 | 8 |
| No plans to internationalize | 16 | 15 | 36 | 42 | 32 |

*Table 10.6    Overseas markets ( per cent )*

| Sales made | Manufacturing only | Manufacturing & KIBS | KIBS, partly ETS | KIBS, all ETS | All |
|---|---|---|---|---|---|
| Western Europe | 71 | 85 | 65 | 68 | 71 |
| Beyond Western Europe | 76 | 65 | 77 | 74 | 75 |
| Central/Eastern Europe | 24 | 15 | 42 | 47 | 33 |
| Beyond Europe | 76 | 60 | 65 | 58 | 66 |
| N | 21 | 20 | 26 | 19 | 92 |

three-quarters were active in markets beyond Western Europe – Central and Eastern Europe providing significant markets, especially for the KIBS firms – and two-thirds were active beyond Europe. These patterns are interesting, because they indicate that, while manufacturers are more likely to be active in Western European markets than are KIBS firms, the difference in participation in markets beyond Western Europe is much less.

To investigate these issues further, we analysed the data using binary logistic regressions where firms with overseas sales are coded one and those without are coded zero. In constructing these, it was anticipated that

1.  Firm size (here measured in terms of the log of the number of employees – Ln(Employment)) would be positively related to participation in overseas markets. Unless their markets are highly concentrated spa-

tially, larger firms tend to serve more widely dispersed markets than small firms.

2. Other things being equal, firms specializing in ETS activities were expected to be more likely to have overseas sales, because of the need to gain access to markets. Non-specialist firms can expand their markets locally by diversifying to meet other local demands – this approach to market expansion is not available to specialist firms. (In the first column of Table 10.7, Y_%ETS captures the percentage of each firm's sales attributed to ETS).

3. Whether the firm belonged to a wider group of firms might have an impact on its internationalization. On the one hand, being part of a group may provide a firm with greater resources to engage in international markets. On the other, there may be a functional division of labour within the group such that international markets are dealt with by other group businesses. It is therefore not clear whether subsidiary status would affect positively or negatively the propensity to participate in overseas markets. (To produce Table 10.7, D_Subsidiary, a value of one is ascribed where respondents have said the NW establishment is a subsidiary of a larger firm or firm group; all others are ascribed zero).

4. The nature of the activity engaged in was expected to have an impact on internationalization. In particular, those firms engaged in manufacturing were expected to be more likely to have international activities than were the KIBS providers (D_KIBS = 1) because for manufacturers the product traded is separable from the means of provision, which is not usually the case with KIBS. For KIBS and other services, the producer must normally be in close proximity to the user or customer, which tends to encourage dispersed local provision. Moreover, KIBS services tend to be less standardized and draw directly on local knowledge to a greater extent than manufacturing production. For all these reasons, we consider that manufacturers are more likely to have overseas sales than are service firms, and KIBS in particular. (To produce Table 10.7, values are ascribed as follows: D_Manufacturing = 1; D_Manuf. & KIBS = 1 if both manufacturing and providing KIBS.)

5. The nature of the demand in markets served may also affect the internationalization of the firms. Some markets, such as the water and energy utilities, are longer established (and standardized) and these may be more internationalized than the newer, less defined markets. Also, if manufactured products are more tradable, then manufacturers may themselves be an important market for the firms, especially the KIBS firms. For example, the KIBS firms in the North West provide (local and foreign) manufacturers with local knowledge about markets, regulations and so on, in order for them to adapt their products for different

*Table 10.7  Logistic regressions for activity in overseas markets.*

| | Any overseas sales (model 1) | Model 1 – reduced | Western Europe (Model 2 – reduced) | Beyond W. Europe (Model 3 – reduced) |
|---|---|---|---|---|
| Ln(Employment) | 0.397*** | 0.379*** | 0.409*** | 0.337*** |
| Y_%ETS | 0.122 | — | — | — |
| D_Subsidiary | −0.032 | — | — | — |
| D_Manufacturing | 1.820** | 1.546*** | 1.444*** | 0.582 (15%) |
| D_Manuf. & KIBS | −0.363 | — | — | — |
| D_KIBS | −1.308 (14%) | −0.779 (17%) | — | −0.806 (13%) |
| D_User_Utilities | 1.145*** | 1.140*** | 1.043*** | 0.740** |
| D_User_Manuf. | −1.100 | — | 0.851** | — |
| D_KIBS for Manuf. | 2.331** | 1.227*** | — | 0.633 (15%) |
| Constant | −0.930 | −1.353** | −2.837*** | −1.262** |
| N | 152 | 152 | 152 | 152 |
| -2LL | 164.8 | 166.5 | 168.8 | 186.5 |
| Model $^2$ (d.f.) | 39.1*** (9) | 37.4*** (5) | 21.2*** (4) | 23.0*** |
| Nagelkerke R$^2$ | 0.307 | 0.295 | 0.302 | 0.187 |

*Notes:*
D_ indicates a dummy variable.
*** indicates significant at 1 per cent.
** indicates significant at 5 per cent.

markets. (To produce Table 10.7, values are ascribed as follows: D_User_Utilities = 1, D_User_Manuf. = 1, D_KIBS for Manuf. = 1).

The models were run three times: first for any overseas sales, second for sales within Western Europe, and third for sales beyond Western Europe. The results, displayed in Table 10.7, show that specialization and ownership status did not have an impact on the propensity to be engaged in international markets. However, as expected, participation in overseas markets increased with firm size, and was greater for manufacturers than for KIBS providers. Those that served the utilities market were also more likely to have international sales. It is interesting to note that those KIBS firms that served manufacturers were more likely to have overseas sales. Unfortunately, we do not know whether the manufacturers the KIBS provided services to were those indigenous to the North West, in which case they may have assisted with the creation of export markets, or whether they were located in other countries, in which case the KIBS would have helped foreign manufacturers serve the North West market.

The model for sales in Western Europe shows that larger firms and manufacturers were more likely to have sales in Western Europe, and those firms whose customers were utilities or manufacturers were also more likely to have sales in Western Europe. Larger firms, manufacturers and firms with utilities customers were also more likely to have sales beyond Western Europe, whilst KIBS, except those with manufacturers as customers, were less likely to have sales beyond Western Europe.

These results point to an important but often hidden role of some KIBS in the internationalization process of a cluster: they can support the development of a cluster by assisting local manufacturers with developing overseas markets, but they can also undermine (at least in the short term) the development of the cluster by opening up the local market to overseas producers, thereby reducing the demand met locally.

Analysed by the mean distribution of their income from 'the North West region', from 'the rest of the UK', and from 'overseas' (we are unable to distinguish between the various overseas regions), the firms in the classifications examined here reveal different distributions of income (see Figure 10.1). At one extreme, the manufacturers without KIBS earned on average only a quarter of their income from the North West region, with half being from the rest of the UK and a quarter from overseas markets. As most of these firms had international sales, the proportion of sales from overseas increased only slightly when the sample was restricted to those with international sales. At the other end of the spectrum, the non-manufacturing KIBS firms dedicated to ETS derived, on average, half their income from within the North West region, with only 9 per cent being from overseas.

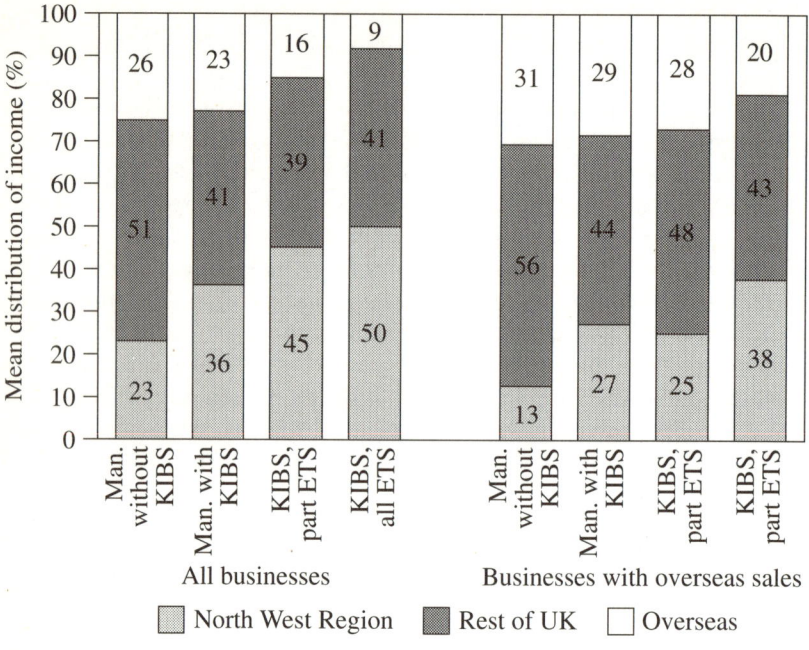

*Figure 10.1　Sales, by spatial reach amongst firms*

Even amongst those that earned some income from overseas, the average proportion was 20 per cent, roughly half that earned from within the North West region. These average patterns, which conceal intra-group variation, suggest manufacturers remain the prime source of overseas earnings, and that the regional trade balance would be improved by strengthening the manufacturing base.

When broken down by the proportion of sales earned in overseas markets (amongst firms with overseas sales) it is apparent that there is greater variance amongst the KIBS providers than amongst the manufacturers (see Table 10.8). Most KIBS with international sales earned only a small proportion of their turnover from those activities. This was also true of many manufacturers. However, some KIBS earned all of their income from overseas activities, indicating that international trade was the main business of these firms. This diversity in sales from international activities points to considerable diversity in the role of KIBS in the North West ETS cluster. Many are very small firms that undertake only occasional international activities, whilst a few are dedicated to international trade as their business.

*Table 10.8     Percentage of turnover from overseas markets (per cent)*

| Sales made overseas | Manufacturing only | Manufacturing & KIBS | KIBS, partly ETS | KIBS, all ETS | All |
|---|---|---|---|---|---|
| Up to 10% | 39 | 40 | 50 | 63 | 51 |
| 11–49% | 33 | 35 | 33 | 21 | 30 |
| 50–99% | 28 | 25 | 4 | 16 | 16 |
| 100% | 0 | 0 | 13 | 0 | 3 |
| N | 18 | 20 | 24 | 19 | 87 |

*Table 10.9     Modalities of international trade (per cent)*

| Sales made overseas | Manufacturing only | Manufacturing & KIBS | KIBS, partly ETS | KIBS, all ETS | All |
|---|---|---|---|---|---|
| Agents/distributors | 71 | 70 | 31 | 42 | 52 |
| Other group businesses | 43 | 35 | 31 | 21 | 32 |
| Joint venture partners | 19 | 30 | 42 | 26 | 30 |
| N | 21 | 20 | 26 | 19 | 92 |

Finally, we examine the means by which firms achieved their overseas sales (see Table 10.9). The most widespread means was by use of agents or distributors, followed by using other offices or businesses in the firm group and joint venture partners. Analysing the results by using logistic regressions (Table 10.10), it is apparent that manufacturers are more likely to use agents and distributors, although subsidiaries are less likely to, probably because internationalization activities are undertaken through other group businesses. Manufacturers, subsidiaries and specialist ETS firms are more likely to internationalize through other group businesses. These patterns suggest that many KIBS with international sales do not rely on a permanent presence in overseas markets through agents, distributors or other group businesses, but serve these markets directly and in person, through personal visits and associates and contacts abroad.

### Towards a Typology of Organizational Structure and Modalities of Internationalization

It is clear from the above discussion that a variety of organizational structures and modalities of operating across national borders exist among the

*Table 10.10  Logistic regressions for modalities of international trade*

| | Agents/distributors (Model 1) | Model 1 – reduced | Through other group offices (Model 2) | Model 2 – reduced |
|---|---|---|---|---|
| Ln(Employment) | 0.188 (14%) | 0.160(14%) | 0.409*** | 0.411 |
| D_Subsidiary | −1.446** | −1.339** | 0.853 (14%) | 0.844 (13%) |
| D_Manufacturing | 1.637*** | 1.447*** | 0.989* | 1.004* |
| D_KIBS | 0.131 | | −0.035 | |
| D_All ETS | 0.526 | | 0.762 (18%) | 0.764 (18%) |
| Constant | −1.112 (16%) | −0.708 (13%) | −3.058*** | −3.091*** |
| N | 92 | 92 | 92 | 92 |
| -2LL | 109.2 | 110.2 | 98.7 | 98.4 |
| Model $x^2$(d.f.) | 18.2*** (5) | 17.1*** (3) | 16.0*** (5) | 16.0*** (4) |
| Nagelkerke $R^2$ | 0.239 | 0.227 | 0.224 | 0.223 |

*Notes:*
D_ indicates a dummy variable.
*** indicates significant at 1 per cent.
** indicates significant at 5 per cent.
* indicates significant at 10 per cent.

ETS firms. Since we have no time-series data we have no way of knowing whether firms move, over time, from one structure/modality to another in a systematic fashion. However, the fact that so many diverse routes to internationalization are evidenced among the firms surveyed suggests that these structures/modalities coexist in a mutually supportive way: that is, different structures are associated with different functions and achieve the outcome of operating across national borders in different ways.

We also do not know whether firms move in and out of these modalities during the course of the development of the firm. Is there a 'point of no return', for example, where a non-internationalized firm becomes an internationalized firm and remains in that state for the rest of the firm's 'life'? Alternatively, could a non-internationalized KIBS firm one year become an 'internationalized KIBS' another, with little systematic planning or forethought (by virtue of a chance meeting of two individuals at a conference, for example), then, once the parameters of a particular contract/relation are complete, resort to being non-internationalized in subsequent years? It seems quite possible that such an idiosyncratic mode of internationalization will be the case sometimes, especially among the large group of small KIBS who are primarily serving local markets, but occasionally and in a limited way venture into overseas markets. Likewise, could a small KIBS be 'born global' by virtue of familial or social ties extending across national borders which are intentionally exploited as part of the strategy of the firm? Again, anecdotal evidence from the survey findings (such questions were not systematically explored on the questionnaire) suggest that this pathway to internationalization is exhibited within the ETS firms of the North West. Such a range of scenarios is somewhat at odds with a view of the internationalization process which sees KIBS firms systematically (and atomistically) moving through 'stages' of internationalization.

In a preliminary attempt to capture this diversity among the internationalized and non-internationalized firms (at the time of the survey) a typology of structures and modalities associated with internationalization (and non-internationalization) is put forward below. Each 'type' is briefly described. These classes are not always mutually exclusive, however. Rather, there are overlaps and firms may belong to two or more of the categories. Conversely, they may move into, or out of, some categories. Further systematic empirical work would be needed to develop, refine and verify the saliency of each of these classifications.

### Internationalized firms
*Ad-hoc internationalization (small KIBS)*. This group includes small autonomous consultancies where the mode of internationalization might be referred to as 'non-strategic'. Turnover from overseas activities constitutes

only a small 'accidental/incremental' percentage of the total turnover of these firms. Managers/proprietors of these micro-firms may have little desire or incentive to develop international markets systematically.

*Strategic internationalization (small KIBS).* This group includes small autonomous units with close familial/social ties to a particular country or countries. A large percentage (in one case from the survey, 100 per cent) of the firms' turnover is derived from contacts/trade with that country. Such firms may therefore be 'born global' (see Toivonen in this volume). This class also includes small KIBS with a specific and systematic overseas development strategy: for example, local consultants advising on land and water remediation in Eastern Europe using European Commission finan-cial incentives (grants to Eastern European accession countries).

*Market intermediaries.* A local base of agent/distributors exists to under-take the very specific and tightly drawn role of importing manufactured products originating overseas and/or opening up international markets to local manufacturers.

*Large multi-establishment/multi-country ETS consultancies.* Multi-specialist international ETS KIBS firms are emerging as a result of acqui-sitions, producing internationally organized multifunctional groups. Business units are organized around specialist divisions of labour, either to provide the group with desired technological competences (computer soft-ware/systems development for example) or to provide access to particular geographical country/region markets (including England's North West). Some North West/UK headquartered firms have become absorbed into these international groups, potentially heralding the onset of a changing global structure of ETS activities with implications for the region. The role of these new 'super' ETS KIBS businesses, particularly in opening up, developing and lubricating markets and coordinating the supply of prod-ucts and services to service international markets, warrants further empir-ical investigation.

*Local subsidiaries (large or small establishments, KIBS or manufacturing, or both).* Local subsidiaries of national or international parents undertake a mix of manufacturing and KIBS activities. Where they sell into local markets this potentially constitutes a leakage from the local economy. Alternatively, their presence may facilitate the development of skills, knowledge and jobs locally. However, balance of payments benefits to the local economy which might accrue from the local subsidiary earning revenue from overseas trade are obscured by intra-firm (non-market) transfers and the fact that income to the firm will be recorded at the registered location of the corporate headquarters.

*Sub-contracting and networks.*   An array of subcontracting arrangements facilitate the carrying out of major 'cradle to grave' overseas projects and these vary from subsector to subsector within ETS. Contexts include major 'turnkey' land or water remediation projects, or the management and operation of effluence treatment plants. Patterns and arrangements of outsourcing (Howells, 2000b) and subcontracting need therefore to be understood in the context of new divisions of labour, and the emergence of new market/exchange interfaces pertinent to individual subsectors of ETS. Weblike structures of interdependency develop to facilitate international trade, which combine firms engaged in consulting (from product development, to laboratory sample testing, to near-to-market customer advisory services, training and marketing) and routine services (parts of recycling, haulage) with manufacturing, assembly, processing and logistics roles. These roles are carried out by a plethora of different firms and firm-types which are coordinated internationally. The survey captured firms involved in these multi-firm arrangements, but the firm survey methodology, especially when territorially bounded, is not capable of capturing the range of interrelationships involved in such networked structures, or their spatial patterning. Further theoretical and empirical work is much needed to identify the extent and role of KIBS involvement, both as main contractors and subcontractors in the coordination and mediation of such networks in different ETS subsectors.

*Vertically integrated manufacturers with in-house service support.*   These include vertically integrated ETS firms manufacturing specialist plant and equipment, often customized to individual client specifications and requirements (for example, emissions monitoring and control systems). As with all capital goods industries, a high level of customer support is provided by highly-skilled specialist staff, with a range of specialist technical knowledge from systems design and new product development to on-site training and customer service. Internationalization occurs through a range of organizational forms and contractual agreements, especially direct supplier–customer contact from an in-house sales force but also including research and development alliances, joint ventures between supplier and client and so on.

**Non-internationalized firms**
*Micro-KIBS who are 'content' to serve regional markets.*   Many local KIBS do not seem motivated to seek out international trade. They are content with a level of earnings provided by consulting for a small number of large

local clients in the private or public sector. Apart from timing and seren-
dipity, there seems little of significance to distinguish this group from those
in the ad-hoc internationalization (small KIBS) group above.

*Large KIBS serving regional markets exclusively.*   This group includes professional practitioners deriving only a small percentage of total turnover
from ETS related work and with no revenue derived from overseas markets.
An example from the survey is a large regional solicitor's practice with a
specialist environmental law team.

*Ancillary ETS.*   As captured in the OECD classification discussed earlier,
this group have only a small percentage of turnover attributable to ETS,
and environmental activities are an ancillary part of the business. Firms in
this group from the survey were not internationalized either in terms of
their primary business activity or, unsurprisingly, in their ancillary ETS
activity. Examples were local establishments of a hotel chain and local
branches of a national bank.

*Not-for-profit/territorially mandated.*   By virtue of restrictions in their
Articles of Association, these organizations, though contributing to the
ETS 'economy' by providing services, training, education, information and
ETS-related skills and jobs, are nonetheless often territorially restricted by
either their funding sponsors (for example, local governments) or their con-
stitution. A key example from the survey is the Groundwork Trust.
Groundwork is an environmental charity with 43 branches in the UK. No
fewer than eight Groundwork branches replied to the ETS survey from
across the North West. In the event, these were excluded from the statistical analysis in order not to obscure the internationalization patterns of
firms. Other non-firm organizations would also come into this category,
such as ETS trade associations and universities, who nonetheless play
various important roles in the functioning of the local ETS 'economy'.

Finally in this section, questions concerning membership of professional and
trade associations among ETS firms are briefly considered. Some possible
links between (1) processes of 'professionalization' of those services, (2) the
development of markets through which to trade those services, and (3)
market expansion through processes of internationalization are put forward.

## Marketization, Professionalization, Internationalization

Elsewhere it has been reported that the occupational group 'environmental
engineers' are 'hardly professionalized' compared to two other KIBS occu-

pations (architects and accountants), but also that environmental engineering is a much newer 'discipline' than the other two (Miles and Boden, 2000). Findings from the ETS survey analysed for this chapter largely concur with this. Respondents were asked to name any national or regional industry or trade association or groups to which they were affiliated. There was a split of around 60:40 between those naming any organization and those naming none – less than 50 per cent of firms with fewer than five employees named one, however. Furthermore, for those who did answer in the affirmative: (1) a huge array of trade associations were put forward, reinforcing the apparent lack of internal cohesion in the activities and affiliations of the sample respondents; (2) many named only very broad-based organizations such as chambers of commerce; (3) the wide range of associations named beyond this were largely connected with the much narrower class of subsectoral specialisms in which the ETS firm operated, such as British Water, the British Nuclear Industries Forum, and the Chemical and Industrial Consultants Association. Only 13 respondents named generic environmental representative bodies such as the Environmental Industries Association. Importantly, many of those who did reply named a range of institutes connected with their 'core' discipline, such as the Institutes of Physics, Mechanical Engineers, Civil Engineers and Chemical Engineers. A number named two organizations, combining a relevant trade association and their own relevant professional institute.

The role of trade associations is to develop inter-firm trade links and to lobby on behalf of an industry's firms. In contrast, professional bodies play an important part in the reproduction and development of embodied sets of 'knowledge' and practices associated with a particular group of skilled workers. A range of strategies are used to achieve this: the establishment of accredited training, conferences, journals, industry awards and, importantly, the setting of industry standards and codes of conduct to regulate members' practices in order to protect and enhance the reputation and trust held in the profession by outsiders. Professions also self-regulate by devising and imposing entry restrictions. These serve to maintain a degree of exclusivity, and have the effect of conserving the status of the profession vis-à-vis others. There is a large body of sociological writing on professionalization projects. However, the role of professional bodies in relation to maintaining and enhancing the economic position of different KIBS groups is not at all well researched.[17]

Elsewhere,[18] our research on the market research 'industry', in particular studying the historical evolution of its 'professional'[19] body, the Market Research Society and that of its close brethren in the marketing and market information industry has begun to uncover related issues concerning the linkages between professionalization, the growth of different KIBS sectors and the development of markets in KIBS sectors. Possible linkages are identified between two sets of strategies. First are strategies used by professional representative bodies to expand their membership (including strategies to recruit international members). Second are strategies to protect the differentiated and exclusive nature of those practices in the face of potential entry and cannibalization from knowledge worker groups with close or associated skills, who have the potential power, therefore, themselves to contact the market researchers' customers with competitive offerings. Additionally, professional bodies appear to play an interesting role in processes of market making and market expansion in knowledge-intensive sectors. By accelerating the commodification of intangibles, these bodies are facilitating their supply and purchase across market interfaces, but at the same time their strategies include attempting to restrict access so that those skills and knowledges do not become known, and traded, by non-members.

Returning to the findings reported in this chapter, it can be seen that the ETS sector has not (yet?) developed the degree of cohesion and self-regulation associated with other 'professional' groups. This has a number of implications. First, there are insufficient procedures and mechanisms to facilitate self-regulation at the level of the 'collective' group of knowledge workers and therefore there is much scope for 'malpractice' both to occur and to pass unpunished. This threatens the level of mutual trust at the buyer–supplier interface and potentially undermines the market exchange process. Second, there is an absence of a representative group able to promote the group and its embodied skills across international borders, thus contributing to the international development of markets for the skills of home-nation workers. But, third, it simply may be that such a professional grouping has not organized itself because the collection of skills and knowledges which comprise the diverse ETS sector cannot achieve a sufficient level of internal cohesion to warrant the establishment of a differentiated 'profession'.

## CONCLUSIONS AND IMPLICATIONS FOR 'CLUSTER' POLICY

This chapter has examined the internationalization of sales among firms active in 'environmental technologies and services' in the North West region of England in the UK. As 'environmental technologies and services' have been identified as one of several growth areas with the potential to create a 'cluster', the development of which is a cornerstone of the North West's regional development strategy, the activities of these firms are worthy of detailed investigation.

The analysis finds that most of the North West ETS firms surveyed are small, low-turnover consultants, which seem to be primarily serving the local needs of manufacturing and public authority clients within the region. The manufacturing firms are more likely to be active in overseas markets than service firms, although KIBS providers seem to play an important role in the internationalization of manufacturers' outputs. They help manufacturers within the North West region gain access to overseas markets, but they also help foreign manufacturers gain access to the UK and North West regional market. In so doing, KIBS play a positive role in supporting 'cluster' development – but can also undermine it.

The 'clusters approach' marks a significant change in regional policy, because the focus is on the cluster as a whole rather than the individual firms that are active within it. Thus the 'clusters approach' can be seen as a variant of the 'systems of innovation' approach to economic development, which encourages the examination of interconnections between organizations rather than treating each organization as an isolated agent. This matters when examining questions such as the extent of internationalization activities, for, while an atomistic approach would take the organization as the unit of analysis, the systems approach would try to examine the cluster as a whole. For example, while a firm may not itself be engaged in international activities, it may play a central role in facilitating the internationalization activities of other firms. A danger with the systems or clusters approach, however, is that it assumes a great deal of interdependence between agents which may not exist and/or which is difficult to demonstrate. Geographical concentrations can arise because of agglomeration economies, but this does not mean the businesses in the agglomeration are necessarily deeply interconnected; often what are labelled 'clusters' are instead agglomerations. It remains an open question to what extent the 'clusters' identified by the NWDA (and other regional development agencies) are 'true clusters', and to what extent they are agglomerations of businesses.

Furthermore, when we consider relationships between 'clusters' (or, more accurately, agglomerations) these may be highly complementary, providing scope for joint 'cluster' initiatives. Membership in alternative clusters may overlap rather than being mutually exclusive. However this also has implications for those vying to drive up 'membership' of individual clusters, where membership requires a significant commitment, in terms of time resources, of being involved. Indeed, a firm representative recruited as a member to one 'cluster' may in fact feel a more salient affiliation, to an alternative 'cluster', so cluster policy may, ironically, result in different cluster 'sponsors' competing against each other for the time and attention of potential 'subscribers'.

What can be stressed, however, is the differentiated nature of the firms within the region, and taking a 'cluster perspective' provides an alternative view to one in which all firms are seen as atomistic competitors, perhaps developing through stages. Such a view neglects much of the diversity amongst the firms, in their functions and trajectories of development. In contrast, it was the diversity of the sample firms in terms of organizational forms and modes of internationalization which was as interesting as the aggregate picture. The variety of organizational modes associated with internationalized KIBS also has implications in terms of questioning the staged-development thesis of internationalization of knowledge-intensive services. Rather, the findings point to the need to understand the scope and significance of micro-diversity and organizational variety. This also has policy implications, pointing to a requirement for a highly segmented and differentiated approach to supporting the internationalization of this particular group of activities. The findings also foreground the scope for organizational innovation associated with the discovery of new and alternative modes for reach and coordination across national frontiers.

## NOTES

1.  See Andersen *et al.* (2000) and Toivonen (2001). Arguably, however, the KIBS acronym overplays the business-to-business dimension of some knowledge-intensive services and plays down the role of the public sector which Howells (2000a, p.274) has called the 'darkest hole' of research on services innovation. In the ETS sector, for example, many 'new' service firms have their origins in spinoffs from the public sector, especially accompanying local government contraction in the UK. Conversely, a high proportion of ETS revenues are derived from trading with local and central government. Indeed, 'KIS' may be a more appropriate acronym than KIBS in this particular context.
2.  The North West of England is a diverse region including the major cities of Manchester and Liverpool, plus the outlying industrial towns in Lancashire and Cheshire (for example, Crewe, Runcorn, Blackburn and Preston) and the largely rural areas of Cheshire, North Lancashire and Cumbria. The NWDA is one of eight private sector-led English regional development agencies established by the UK government responsible to the (then) Department of the Environment, Transport and the Regions. The RDAs have been operational since April 1999.
3.  These are 'aerospace', 'automotive', 'chemicals', 'energy', 'food and drink', 'mechanical and other engineering (including marine)' and 'textiles'.
4.  These are 'computer software and services (ICT)', 'creative industries (including media)', 'environmental technologies and services', 'financial and professional services', 'life sciences' (biotech, pharmaceuticals and health), 'medical equipment and technology' and 'tourism'.
5.  A notion discussed in more detail in the next section.
6.  Also notable is that the NWDA uses the term 'sector' interchangeably with 'cluster'.
7.  Miles (2000) developed a 'knowledge-based' classification system of ETS firms using cluster analysis to identify 'knowledge-sets' which differentiated around 100 UK ETS firms. He found that, although very different service functions were undertaken by different KIBS firms, there was a high degree of specialization by KIBS subsectors, and little overlap between subsectors in terms of the knowledge-sets of firms This finding of a

high degree of subsector exclusivity in ETS firms is broadly consistent with the findings reported in this chapter.

8. The definition of a science is of course not fixed, and subdisciplines of 'old sciences' can 'spin out' to become autonomous 'new sciences'.

9. Here the word 'demand' is used with caution. Elsewhere, we discuss in greater detail a theoretical position concerned with understanding the 'instituted' nature of exchange relations, both market and non-market, which sheds quite a different light on understandings of how the 'production' of goods and services comes to articulate with their 'consumption' (Harvey *et al.*, 2001).

10. Established as a Company Limited by Guarantee with funding from the Business Development Fund under the Cluster Development Strategy of the NWDA .

11. See *http://www.envirolinknw.co.uk/*

12. This embraced 12 environmental technology and service 'areas': air pollution control; contaminated land remediation; energy management; environmental monitoring and instrumentation; environmental services; landscape (services); marine pollution; noise and vibration control; renewable energy; transport pollution; waste management; and water and wastewater treatment. Each of these was outlined in the questionnaire, with a brief description of the 'area'.

13. These data have previously been analysed for Envirolink by Business and Market Research Ltd, and the ETS consultants EnvirosMarch provided a summary report of key findings for regional policy makers in August 2000.

14. These being defined by the 12 'areas' listed in note 12, which were listed and briefly defined in the questionnaire.

15. The extent of sales in ETS may also be related to the unit of analysis. For instance, if the response was from a business unit within a wider firm group, the proportion of sales attributed to ETS might be higher than for the firm as a whole.

16. Including one respondent in software production.

17. Miles (2001) provides a very preliminary entry point into such a research agenda.

18. Warde, Randles and McMeekin (2001).

19. We use the word reservedly, since some would contest the use of the term 'profession' to describe marketing services.

# REFERENCES

Andersen, B., J.S. Metcalfe and B. Tether (2000), 'Distributed Innovation Systems and Instituted Economic Process', in J.S. Metcalfe and I. Miles (eds), *Innovation Systems in the Service Economy*, Boston: Kluwer.

Andersen, B., J. Howells, R. Hull, I. Miles and J. Roberts (eds) (2000), *Knowledge and Innovation in the New Service Economy*, Cheltenham, UK and Northampton, MA, USA: Edward Elgar.

Commission of the European Communities (CEC) (1997), 'Building a Sustainable Europe: Communication from the Commission on Environment and Employment', CEC, COM(97)592final, Brussels.

Harvey, M., McMeekin, A., Randles, S., Southerton, D., Tether, B. and A. Warde (2001), 'Between Demand and Consumption: A Framework for Research', CRIC Discussion Paper no. 40, CRIC, University of Manchester and UMIST.

Howells, J. (2000a),'Understanding the New Service Economy', in B. Andersen, J. Howells, R. Hull, I. Miles and J. Roberts (eds), *Knowledge and Innovation in the New Service Economy*, Cheltenham, UK and Northampton, MA, USA: Edward Elgar.

Howells, J. (2000b), 'Outsourcing Novelty: The Externalisation of Innovative

Activity', in B. Andersen, J. Howells, R. Hull, I. Miles and J. Roberts (eds), *Knowledge and Innovation in the New Service Economy*, Cheltenham, UK and Northampton, MA, USA: Edward Elgar.

Miles, I. (2000), 'Environmental Services: Sustaining Knowledge', in B. Andersen, J. Howells, R. Hull, I. Miles and J. Roberts (eds), *Knowledge and Innovation in the New Service Economy*, Cheltenham, UK and Northampton, MA, USA: Edward Elgar.

Miles, I. (2001), 'Taking the pulse of the knowledge-driven economy: The role of KIBS', in M. Toivonen, M. (ed.), *Growth and Significance of Knowledge Intensive Business Services (KIBS)*, Helsinki: Employment and Economic Development Centre for Uusimaa.

NWDA (2000), 'North West Innovation Strategy', North West Development Agency.

OECD (1993), *Pollution Abatement and Control Expenditures in OECD Countries*, Environment Monograph no. 38, Paris: OECD.

OECD (1996a), *The Global Environmental Goods and Services Industry*, Paris: OECD.

OECD (1996b), *The Environment Industry: The Washington Meeting*, Paris: OECD.

OECD (1996c), *Interim Definition and Classification of the Environment Industry*, OECD/GD(96)117, Paris: OECD.

OECD (1999), *Mapping the Environmental Goods and Services Industry*, Directorate for Science, Technology and Industry, Paris: OECD.

Toivonen, M. (ed.) (2001), *Growth and Significance of Knowledge Intensive Business Services (KIBS)*, Helsinki: Employment and Economic Development Centre for Uusimaa.

Warde, A., S. Randles and A. McMeekin (2001), 'Market Research and Market Formation: Professionalization, Institutionalization, Structure and Dynamics in the Construction of Markets for Market Research', paper presented at the IIDE-CEPN/CRIC workshop, 'The Market and the Organisation of Exchange', Paris, November.

# Index

accounting services 215, 216, 218, 220
acquisitions (takeovers) 69, 140, 215
advertising 99, 215
Aharoni, Y. 141, 162, 166, 169
airlines 92–3
Allen, J. 122
American Airlines 92–3
American Express 93
Andersen Consulting 177
Antonelli, C. 176
AOL 173
Arrow, K. J. 171
assets, intangible 17
assimilationism 21, 25
Association of Southeast Asian
        Nations (ASEAN) 91
auditing 215, 218
Austria
    business services 142, 152, 153–4, 155
        regional concentration 143, 152,
            153–4, 155
auto industry 90–91

Baark, E. 177
Bagchi-Sen, S. 162
Baker, P. 17
balance of payments, data on 84
banking 93
Barbados 92–3
Bargas, S. E. 214
Barras, R. 20, 51
barriers to internationalization of
        services 196–202, 203–4, 218–20
Bartlett, C. A. 120
Belgium
    business services 142, 143, 146, 152,
        155
        regional concentration 143, 152,
            155
Bélis-Bourgouignan, M.-C. 119
Belleflamme, C. 20
Bhagwati, J. N. 35, 36, 38

Birkinshaw, J. 177
Boddewyn, J. J. 166
Bonamy, J. 140
Brazil 41
Breathnach, P. 173
Brynjolfsson, E. 38
business services
    externalization/outsourcing 2, 18,
        94–5, 162
    foreign direct investment (FDI)
        139–40, 163–5
    globalization and 138–40
    growth in 2
    international trade 163
    internationalization of UK firms
        161–80
        factors influencing pattern of
            168–71
        findings of study 166–72
        research strategy 165
        resource-oriented activity 172–8
        stages 166–8
    regional concentration of innovative
        services 137–56
        globalization and 138–40
        implications for regional
            development 140–42
        overview of European regions
            142–51
        regional profile of European
            regions by per capita income
            152–4

Campbell, A. 17
Canada 22
Cantwell, J. 19, 43
car industry 90–1
Castells, M. 22, 211
Catalytic Software 93
Cavanagh, J. 18
Chadwick, M. 26, 27, 64
Chesnais, F. 19

Cisco Systems 93
Clairmonte, F. 18
Clark, C. 36
Clegg, J. 34
cluster policy 227–8, 250–52
Coase, R. H. 38
Coe, N. M. 123
Compaq 126
competition 25, 59, 69, 190, 218
  competitive advantages 17
  coordination 219
  EU policy 60, 61
  for location of
      multinational/transnational
      corporations 19, 28
  regulation of 60
competitiveness, internationalization of
      services and 67–70
computer software 64
consulting services 215–16, 219, 235
Coombs, R. 20
coterminality, services and 64
Cowling, K. 43
Cox, K. 122
Cuadrado-Roura, J. R. 140
Curtis, J. 173
customer care and support services 173
customization 62

Dalum, B. 38
Daniels, P. W. 27, 118, 126, 140, 219
data processing services 99
De Bandt, J. 140
Deardoff, A. V. 68
delivery 20, 203
Dell 130
demarcation between services and
      manufactures 33–52
  internationalization and 41–4
    indices 44–5
    results of studies of
      multinational/transnational
      corporations 45–50
  new international division of labour
      and 39–41
  productivity criterion 36–9
  tangibility criterion 34–6
Denmark
  internationalization of services in
      184–204

barriers 196–202, 203–4
  characteristics of service
      internationalization 187–9
  potentials 189–90
  service characteristics 192–6
  size of firms 190–92
  survey design 185–7
deregulation 24, 25, 28
Dicken, P. 17, 18, 41, 117, 119, 120,
      121, 126, 129
division of labour, international 18,
      39–41, 141
downsizing 16
Dunning, J. 16, 19, 42, 63, 88, 126, 162,
      168–9
Dutka, A. B. 178

economic growth 1
  internationalization of services and
      67–70
EDS 162
embodying of services in goods 64
Enderwick, P. 15, 16, 17, 34, 162
engineering services 220
Envirolink 233–7
environmental technologies firms
      227–52
  definition of environmental
      technologies 229–33
  Envirolink and 233–7
  internationalization 237–43, 245–7
    marketing and professionalization
      and 248–50
    modalities 243–5
    non-internationalized firms 247–8
Esperança, J.-P. 162
European Union (EU) 3, 25, 216
  competition policy 60, 61
  foreign direct investment (FDI) in
      74, 77–8
  integration of services in 75–81
  internal market 60–61, 67
  international trade and 70, 71, 72–3
  internationalization of services in
      59–82
  regional concentration of innovative
      services 137, 139–40, 142–56
Evangelista, R. 21, 22
externalization (outsourcing) 2, 18,
      94–5, 162

Fagan, R. 121, 130
financial services 17, 93, 101, 195, 218
  accounting 215, 216, 218, 220
  auditing 215, 218
  banking 93
  insurance 92, 101
Findlay, C. 178
Finland
  business services 142, 146, 152, 153,
    154, 155, 206–24
    consequences and obstacles in
      internationalization process
      218–20
    degree of internationalization
      208–10
    forms of internationalization
      210–17
    qualification requirements 220–23
    regional concentration 152, 153,
      154, 155
Fisher, A. G. B. 36
Fontagné, L. 66
Ford Motor Co 90–91
foreign affiliates 65, 89, 214–15, 246–7
  data on 78, 85–6, 96–7, 99
  integrated international production
    and 89–90, 102, 119–20
  intra-firm trade 95, 101–9
foreign direct investment (FDI)
  business services 139–40, 163–5
  data on 85
  growth/competitiveness and 69
  integration of services and 77–8
  international trade and 66–7,
    95–101, 110–11
  in services 27, 73–4, 95–101, 110–11
  US IT investment in SE Asia 117–34
    conceptualization of international
      business and 119–22
    dimensions of spatial variation
      129–32
    methodology of study 122–3
    profiling 123–8
Forsgren, M. 126
France
  service sector in 23, 142, 146, 152,
    154
    international trade and 163
    regional concentration 143, 152,
      154

franchising 27, 65, 166
François, J. T. 68
Freeman, C. 18
Frobel, F. 40

Gadrey, J. 140, 175, 176
Gago-Saldaña, D. 151
Gallouj, F. 175, 176, 214
General Agreement on Trade in
    Services (GATS) 25, 26, 60, 178
General Electric Corporation 93
Gentle, C. 177
Germany 22, 23
Ghoshal, S. 120
Global Services Network 25
globalization
  innovative business services and
    138–40
  regionalization and 120–21, 126, 137
  services and 1, 19
Goedegbuurte, R. V. 63
Goodman, D. 130
Gordon, R. J. 37
government and the state
  internationalization of services and
    24–8
  multinational/transnational
    corporations and 19
  *see also* policies
Gray, H. P. 35
Greenwood, R. 118
Grosse, R. 141, 169, 175, 176
Groundwork Trust 248
Groupware 172
Grubel, H. G. 34, 35

Hatzichronoglou, T. 62
headquarters 122, 126–8
Hedlund, G. 166
Helpman, E. 22
Hermelin, B. 208
Hewlett Packard (HP) 123, 126
Hill, T. P. 34
Hindley, B. 68
Hipp, C. 22, 219
Hitt, L. M. 38
Ho, K. 122, 126
Hoekman, B. M. 34
Honda 91
Hong Kong and Shanghai Bank 93

Hood, N. 95, 166
Howells, J. 18, 51, 119, 177, 178, 206,
    214, 215, 218, 220, 222
Hymer, S. H. 43

IBM 123, 126
Ietto-Gillies, G. 42, 43, 44
Illeris, S. 140, 141
India 41, 92, 93, 129, 173
information and communication
    technologies (ICT) 2–3, 15, 18,
    33, 203
  breakdown of demarcation between
    services and manufacturing and
    36, 37
  internationalization of services and
    60, 61–3, 93
  new international division of labour
    and 40–41, 141
  productivity and 37–9
  resource-oriented international
    service activity and 172, 173,
    175–8
  service sector innovation and 20
  US IT investment in SE Asia 117–34
    conceptualization of international
      business and 119–22
    dimensions of spatial variation
      129–32
    methodology of study 122–3
    profiling 123–8
Informix 132
Infosys Technologies 93
innovation
  internationalization of services and
    15–29
  national systems of 18–19, 22, 23–4
  in services 15–16, 19–24, 29, 35–6
  services and 2
    regional concentration of
      innovative services 137–56
institutions, standardization 18
insurance 92, 101
intangible assets 17
integration
  conceptualization of international
    business and 119–22
  EU service sector 75–81
  international production 15, 16–19,
    42–3, 89–95, 101–9, 119–20

of services 77–8
vertical integration 247
Intel 93
intellectual property rights 21
interaction, services and 64
internalization advantages 170–71
international division of labour 18,
    39–41, 141
international production 15, 16–19,
    42–3, 89–95, 101–9, 119–20
international trade
  business services 163, 166–7
  data on 70–3, 84
  foreign direct investment (FDI) and
    66–7, 95–101, 110–11
  information and communication
    technologies (ICT) and 2
  liberalization 3, 25
  modes of 64–6
internationalization of manufacturing
    41–50, 166, 210–11
internationalization of services 1, 3–5,
    117–18
  barriers to 196–202, 203–4, 218–20
  data on transactions and 59–82
  demarcation between services and
    manufactures and 41–4
    indices 44–5
    results of studies of
      multinational/transnational
      corporations 45–50
  Denmark study 184–204
    barriers 196–202, 203–4
    characteristics of service
      internationalization 187–9
    potentials 189–90
    service characteristics 192–6
    size of firms 190–92
    survey design 185–7
  environmental technologies firms in
    NW England 237–43, 245–7
    marketing and professionalization
      and 248–50
    modalities 243–5
  factors promoting 60–63
  Finland 206–24
    consequences and obstacles in
      process 218–20
    degree of internationalization
      208–10

forms of internationalization
210–17
qualification requirements 220–23
growth/competitiveness and 67–70
innovation and 15–29
modes of 87–111, 243–5
evidence from data 95–109
reasons 88–95
policy changes and 24–8
study of UK business service firms
161–80
factors influencing pattern of
internationalization 168–71
findings 166–72
research strategy 165
resource-oriented activity 172–8
stages of internationalization
166–8
types of service transactions 63–7
Internet 138–9, 172, 219
intra-firm trade 95, 101–9
investment 27
liberalization 3, 25
Multilateral Agreement on
Investment (MAI) 26
*see also* foreign direct investment
(FDI)
Ireland 41, 92, 129, 173

Japan
international trade and 70, 71
manufacturing sector in 22
multinational/transnational
corporations in 45
service sector in 22–3
Johanson, J. 166
John, R. 41
joint ventures 27, 66, 166

Kaldor, N. 36
Karsenty, G. 64, 65
Kirkpatrick, C. 126
Kitson, M. 36
Kravis, B. I. 92
Krugman, P. 22
Kverneland, A. 166

labour market 217
Lakha, S. 173
Landesmann, M. 22, 24

Larsen, J. N. 175
legal services 215
Lévy, B. 121
liberalization
international trade 3, 25
investment 3, 25
public services 26
licensing 27, 65
Lipsey, R. E. 69, 92
locational advantages 169–70
Lucas, N. 173
Lundvall, B. Å. 18

McKinsey Co 162, 177
Malaysia 130
Mallampally, P. 15, 17, 25, 41
Mann, M. A. 214
manufacturing sector 230
demarcation with services 33–52
internationalization and 41–50
new international division of
labour and 39–41
productivity criterion 36–9
tangibility criterion 34–6
industrialization period 138
innovation in 20, 21, 22
internationalization and 41–50, 166,
210–11
production 162
international 16–17, 42–3, 89–95,
119–20
Markusen, A. 117
Marshall, J. N. 173
Marshall, N. 140, 151
Martinelli, F. 18, 22, 24, 208
Mason, G. 24
materiality (tangibility), criterion for
demarcation between services
and manufactures 34–6, 63
Mathews, J. A. 123
Maula, M. 218
Mercosur 25, 60
mergers and acquisitions 69, 140, 215
Michie, J. 36
Microsoft 126, 130, 131
Miles, I. 15, 16, 17, 20, 22, 36, 176, 206,
219, 249
Miozzo, M. 18, 21, 29, 51, 62, 88, 96,
172
Mirza, H. 120

mode of presence 27
Moore, K. 177
Morocco 93
Moulaert, F. 140, 219
Multilateral Agreement on Investment
    (MAI) 26
multinational/transnational
    corporations 2–3, 33, 87
  changing modes of
    internationalization of services
    and 87–111
    evidence from data 95–109
    reasons 88–95
  competition for location 19, 28
  conceptualization of international
    business 119–22
  international production 15, 16–19,
    89–95, 101–9, 119–20
  internationalization of services and
    45–50
  intra-firm trade 95, 101–9
  regional headquarters 122, 126–8
  technology and 19
  *see also* foreign affiliates; foreign
    direct investment (FDI)

Nachum, L. 162
national systems of innovation 18–19,
    22, 23–4
Nayyar, D. 27, 36
Nelson, R. 18
Nestlé 130
networking 217, 247
New York Life 92
North American Free Trade
    Agreement (NAFTA) 25, 60
Noyelle, T. J. 178

O'Farrell, P. N. 162
Oliner, S. D. 37–8
Oracle 126, 130
Organization for Economic
    Cooperation and Development
    (OECD) 230–31, 232
outsourcing (externalization) 2, 18,
    94–5, 162
ownership advantages 169

Palmer, R. 42
Paquette, P. C. 173

Patel, P. 19
Pavitt, K. 19, 21
Pearce, R. 42
Peneder, M. 69
Peoples, J. 43
Perry, M. 18, 127, 128, 162
Peters, E. 95
Petit, P. 22, 24, 140
Philippines 173
policies
  cluster policy in UK 227–8, 250–52
  innovation and 22
  internationalization of services and
    changes in 24–8
Pollett, C. 18
Poon, J. 120, 121, 127, 128
post-industrial society theories 15
Preissl, B. 39, 214
PricewaterhouseCoopers 162
Pritchard, B. 121, 130
privatization 26
process innovations 20, 21
product innovations 21
production
  international 15, 16–19, 42–3, 89–95,
    101–9, 119–20
  national systems of 18–19
  services 35
productivity, criterion for demarcation
    between services and
    manufactures 36–9
public sector 26, 232

qualification requirements,
    internationalization of services
    and 220–23
Quinn, J. B. 173

reciprocal arrangements 166
regional development policy 227–8,
    250–52
regionalization 120–21, 126
  concentration of innovative services
    137–56
  globalization and 138–40
  implications for regional
    development 140–42
  overview of European regions
    142–51
  regional profile of European

regions by per capita income 152–4
regulation
  of competition 60
  reforms 60
    deregulation 24, 25, 28
  variations in 42–3, 172
Reich, R. B. 121
research, agenda for 28–9, 133
research and development (R&D) 22, 126
resource-oriented international service activity 172–8
reverse product cycle 20
Richardson, R. 173
Riddle, Dorothy 17, 140
Roberts, J. 17, 34, 140, 165, 175, 176, 178, 210, 215, 216, 217, 218, 219
Rönkkö, P. 218
Rubalcaba-Bermejo, L. 62, 139, 140, 142, 143
Rugman, A. 120
Ruigrok, W. 119

Sampson, G. P. 64, 68
Sapir, A. 34, 64
Sauvant, K. 15, 17, 96
Saxenian, A. 217
scale economies 18
Schienstock, G. 211
scope economies 18
Sen, J. 162
servuction approach 20
Shrimpton, M. 18
Sichel, D. E. 37–8
Silvestrou, R. 22
Singapore
  financial services 41, 93
  foreign direct investment (FDI) in 118
  US IT investment 122–34
Sirilli, G. 21
Smith, A. 68
Smith, N. 122
Snape, R. H. 64, 68
Soete, L. 18, 21, 29, 51, 62, 88, 96, 172
Solow, Robert 37
Sowels, N. 166
spatial variation dimensions 129–32
  interregional 129–30

intra-activity variations 132
  intraregional variations 130–31
spillovers 69–70
Stallings, B. 121
standardization 18, 62, 202
statistical data sources 67, 84–6
Stern, R. M. 34
Stevens, G. 69
Stigler, G. J. 38, 171
Stopford, J. M. 166
Strambach, S. 121, 217
sub-contracting 247
Sugden, R. 43
Sullivan, D. 42
Sun Microsystems 129
Sundbo, J. 173, 214
Swissair 92
Swyngedouw, E. 122

takeovers 69, 140, 215
tangibility, criterion for demarcation between services and manufactures 34–6, 63
technical services 215
technology
  changing modes of internationalization of services and 88–9, 110
  competitive advantages 17
  multinational/transnational corporations and 19
  services and technological change 1, 2
  spillovers 69–70
  standardization 18
  *see also* environmental technologies firms; information and communication technologies (ICT)
telecommunications 3, 93
Telefonica (Spain) 93
Tether, B. S. 22
Toivonen, M. I. 206
Tomlinson, M. 17, 22, 36, 37
Tordoir, P. P. 222
Toyota 91
trade *see* international trade
transnational corporations *see* multinational/transnational corporations
Turnbull, P. W. 166

United Kingdom
  cluster policy 227–8, 250–52
  multinational/transnational
      corporations in 45–6
  new international division of labour
      and 41
  service sector in 22–3, 77, 142, 143,
      146, 152, 153, 154, 155
    environmental technologies firms
      227–52
    international trade and 163
    internationalization of 161–80
    regional concentration 143, 152,
      153, 154, 155
United Nations
  Conference on Trade and
      Development (UNCTAD)
      17–18, 45, 87, 88, 90, 91, 92,
      102, 118, 119, 128, 163
  CTC 162
United States of America
  foreign direct investment (FDI) by
      67
    international trade and
      95–101
    investment in SE Asia 117–34
  international trade 70–71, 163
    FDI and 95–101
  multinational/transnational
      corporations in 45
  service sector in 22, 23
    international trade and
      163

Vaitsos, C. 18
Valeyre, A. 140
van Tulder, R. 119
Vandermerwe, S. 26, 27, 64
Verbeke, A. 17
Verdoon, P. J. 36
vertical integration 247

Wagner, K. 24
Wallace, P. 173
Warf, B. 18
Warren, T. 178
Weinstein, A. K. 162
Wells, L. T. 166
Werner, R. 214
Wiedersheim-Paul, F. 166
Williamson, O. E. 38
Winter, C. 34, 64
Wong, P. K. 123
Wood, M. 119
Wood, P. 140, 142, 151, 162
World Bank 25
World Development Movement 26
World Trade Organization (WTO) 25,
      26, 60, 67, 178, 184
WPP Group 162
Wymbs, C. 19

Yeung, H. 120, 121, 122, 127, 128
Young, S. 166

Zeile, W. J. 107
Zimny, Z. 25, 41, 96